Emergency Exercise Development

World Health Organization
Western Pacific Region

WHO Library Cataloguing in Publication Data

Emergency exercise development.

1. Emergencies. 2. Disaster preparedness.

ISBN 978 92 9061 419 7 (NLM Classification:WA 295)

FOREWORD

The following document was prepared by the World Health Organization. This document on Emergency Exercise Development was originally designed, produced and distributed by the Federal Emergency Management Agency, Emergency Management Institute of the United States of America. The World Health Organization has added to it and reformatted many sections of this material in order for the users to focus on public health emergencies.

These guidelines are intended to provide a wide range of information related to public health emergency management exercise development. Users should be aware that this document is not regulatory and represents guidelines only. It should not replace appropriate consultation with public health emergency management professionals. The contributors are not responsible for use or adaptation issues arising from the use of these guidelines. Adaptation of this framework is the sole responsibility of the users.

This course material has been designed for emergency exercise staff to acquire in-depth knowledge and skills related to the exercise development process, including management, control, simulation and evaluation. Emphasis is on the construction of exercise planning documents; staffing and training of team leaders in control, simulation and evaluation; the development of expected player actions and points of review; and exercise administration and logistics.

For the novice level exercise planner, it is suggested that they begin by understanding the guidance provided by WHO at http://www.wpro.who.int/NR/rdonlyres/DA340E3E-D27E-47A6-9833-452E7AAC9ED5/0/EDTedDRAFT1ExerciseDevelopmentGuide.pdf

This document and the tools included are in no way comprehensive of all the elements required in the creation and facilitation of an emergency exercise. As areas for improvement are discovered and revisions made, both this guide and emergency plans will change. A comprehensive emergency exercise programme should be part of an overall emergency management programme. Through the cycle of plan development, training, exercising your plan, evaluation, and revising your plan, operational readiness improves and you will more effectively meet the challenges that you and your organization may face in times of emergency.

TABLE OF CONTENTS

Page

Page

Unit 4: Exercise Design Steps

Unit 5: The Tabletop Exercise

Page

Page

Page

Course Overview

About this course

Emergencies can be limited in scope, or reach disaster proportions, sweeping through entire communities, nations and around the world, creating a pandemic. Being prepared to respond to and recover from a pandemic is everyone's challenge. Whether your organization is a government agency tasked with a particular response role, a volunteer agency that responds to community needs, or a private sector entity that may be faced with a pandemic situation, you have an important role in preparation.

As an outcome of your organization's emergency planning process, plans should be in place that specify how you prepare for a pandemic or potential communicable disease events that may have international consequences, how you will respond if an event occurs, how you will mitigate the potential effects of emergencies, and how you will recover. Practice is an important aspect of the preparation process.

Experience and data show that exercises are a practical, efficient, and cost-effective way for organizations in the government, non-profit and private sectors to prepare for emergency response and recovery.

This course is based on one important premise: emergency exercises are worth the effort. Exercises identify areas of proficiency and those that need improvement. Lessons learnt from exercises can be used to revise operational plans and provide a basis for training to improve proficiency in executing those plans.

This course is designed to introduce you to the fundamentals of exercise design and to prepare you to design and conduct an exercise for your organization. It addresses:

- the value of conducting exercises;
- the components of a comprehensive exercise programme;
- the exercise development process—development tasks, organization of the design team, exercise documentation, and the steps in designing an exercise.

This course will cover the purpose, characteristics and requirements of three main types of exercises:

- tabletop exercise
- functional exercise
- full-scale exercise.

In addition this course will cover:

- exercise evaluation
- exercise enhancements
- designing a functional exercise.

Course prerequisites

Exercise design has no prerequisites.

Course completion

Upon completion of this course, you should be able to deliver a basic in-house Emergency Exercise Training Course. You can tailor the content to meet the skill level of your participants, but you are encouraged to include as much information as possible.

This course is designed so that you can develop a solid understanding of the art and science of exercise development. We will take a short break after most units, to allow time to think about the material, discuss it with other participants, and particularly, how it relates to your work and the exercise design function of emergency management.

Exercise design has 11 units. Each of these units is described below.

- **Unit 1: Introduction to Exercise Design,** explores the benefits organizations derive from exercising and leads you through a preliminary needs assessment.
- **Unit 2: Comprehensive Exercise Programme,** provides an overview of five main types of exercise activities that make up a comprehensive exercise programme.
- **Unit 3: The Exercise Process,** presents an overview of the process used to plan and implement a single exercise within a comprehensive programme. It also introduces four key design documents. In this unit, you will take some preliminary "groundwork" steps, including a resource self-assessment and identifying potential design team members.
- **Unit 4: Exercise Design Steps,** takes you through an eight-step process for designing an exercise and provides a variety of job aids to use in applying that process. You will practice each step in an abbreviated fashion as you progress through the unit.
- **Unit 5: The Tabletop Exercise,** takes a closer look at the tabletop exercise, including key characteristics, best uses, guidelines for facilitation and special design considerations.
- **Unit 6: The Functional Exercise,** examines the functional exercise more closely, again focusing on key characteristics, best uses, guidelines for control and special design considerations.
- **Unit 7: The Full-Scale Exercise,** takes a similar look at full-scale exercises and how they differ from the other types of exercises.
- **Unit 8: Exercise Evaluation,** briefly discusses key aspects of evaluation methodology and the evaluation tasks that must take place before, during and after an exercise.
- **Unit 9: Exercise Enhancements,** presents ideas for enhancing an exercise through visuals, equipment, props and people.

About this course (continued)

- **Unit 10: Designing a Functional Exercise,** walks you through applying the eight design steps, using the provided job aids, in developing a simple functional exercise.

- **Unit 11: Course Summary,** prepares you for the final exam by presenting a brief review of the key points covered in the course.

Activities

This course will involve you actively as a learner by including activities that highlight basic concepts. These activities emphasize different learning points, so be sure to complete all of them. Compare your answers to the answers provided following each activity. If any of your answers is incorrect, go back and review the material and or ask the instructor for assistance.

Knowledge checks

Units 1 through 10 are followed by a knowledge check that asks you to answer questions that pertain the unit content. Answers are provided following each knowledge check. When you finish each knowledge check, compare your answers to those provided, and review the parts of the text that you do not understand. It is to your benefit to ensure that you have mastered the current unit before proceeding to the next unit.

Exercise design tool box

A tool box is also included with the course materials. The exercise design tool box contains a variety of exercise examples and templates to supplement the instruction. The course materials will instruct you when to use the tool box.

Job aids

Throughout the course, you will find job aids designed to supplement the text and interactive activities. You can use job aids during the course, and you will find them useful later, after you have completed the course. Copies of all job aids are assembled in Appendix A for easy reference.

Appendixes

At the back of this course are appendixes that include, in addition to the job aids, an acronym list.

Unit 1: Introduction to Exercise Design

Introduction

An **exercise** is a focused practice activity that places participants in a simulated situation requiring them to function in the capacity that would be expected of them in a real event. Its purpose is to promote preparedness by testing policies and plans and training personnel. In this unit, you will explore some of the benefits that organizations derive from exercising.

Unit 1 objectives

After completing this unit, you should be able to:

- explain the value of exercises in improving the four phases of comprehensive emergency management;
- identify the major reasons for conducting exercises.

Why exercise?

Exercises are conducted to evaluate an organization's capability to execute one or more portions of its response plan or contingency plan. Many successful responses to emergencies have demonstrated that exercising pays huge dividends when an emergency occurs.

Why Exercise? (Continued)

Reasons to exercise

The goal in exercise design is to establish a comprehensive exercise programme, one based on a long-term, carefully constructed plan. In a comprehensive programme, exercises build upon one another to meet specific operational goals. The aim is to provide competence in all emergency functions.

There are two main benefits of an exercise programme:

- **Individual training:** exercising enables people to practice their roles and gain experience.

- **System improvement:** exercising improves the organization's system for managing emergencies.

These benefits arise not just from exercising, but from evaluating the exercise and acting upon the recommendations. An exercise has value only when it leads to improvement.

To summarize, there are a number of reasons to perform exercises. Through exercises, organizations can:

- test and evaluate plans, policies, and procedures
- reveal planning weaknesses
- reveal gaps in resources
- improve organizational coordination and communications
- clarify roles and responsibilities
- train personnel in roles and responsibilities
- improve individual performance
- gain programme recognition and support of officials
- satisfy regulatory requirements.

The focus of an exercise should always be on locating and eliminating problems **before** an actual emergency occurs. Corrective actions are an important part of exercise design, evaluation and follow up.

Regulatory requirements

Because the human and monetary costs of emergencies and disasters are so high, governments, agencies, and many corporate entities and other governing bodies have mandated preparedness training and exercising. Several examples of mandated exercises are listed below.

- State and local governments receiving federal funds may have to comply with certain exercise requirements. The Federal Emergency Management Agency (FEMA) in the United States of America has requirements that change periodically, but the programme is normally structured around a four-year cycle.

- Nuclear power plants must exercise their plans yearly, conducting a full-scale exercise every two years. This exercise is evaluated by the United States Nuclear Regulatory Commission (NRC).

- Airports, hospitals, and other health care facilities must conduct a full-scale exercise once every two years to maintain their certification or license to operate.

- Many employers are required by the United States Occupational Safety and Health Administration (OSHA) to develop an emergency action plan and the agency recommends that such plans be exercised at least annually.

- Your government may also have specific regulatory requirements mandating emergency exercises. Note your obligations here:

1. ___________________
2. ___________________
3. ___________________

Functions

In planning exercises, the emphasis is on functions rather than on types of emergencies, because preparedness in those functions is common to all emergencies. **Functions** are actions or operations required in emergency response or recovery.

Many emergency management agencies note 13 functions in their Emergency Management Exercise Reporting System.

- Alert notification (emergency response)
- Warning (public)
- Communications
- Coordination and control
- Emergency public information
- Damage assessment
- Health and medical
- Individual/family assistance
- Public safety
- Public works/engineering
- Transportation
- Resource management
- Continuity of government

Functions (Continued)

If your organization is a private or volunteer organization, you may find it practical to exercise a somewhat different or more limited set of functions. Each of the functions listed above has a set of sub-functions related to it, and your group may focus on some of those. For example, your emergency response focus may relate to efforts such as:

- the management and distribution of donations;
- the logistics of providing needed resources;
- the temporary conversion of a manufacturing process to provide emergency supplies;
- how to coordinate with other organizations to provide mass care;
- how your employees respond to an internal emergency.

The variations are, of course, endless. However your entity is organized, the point is that your exercise programme should identify the applicable functions and emphasize testing the operational procedures within those functions, regardless of the type of emergency.

Activity: needs assessment

Below is a job aid to help you assess your organization's exercise needs. Use this tool to analyze where you may wish to focus your organization's exercise design efforts. In completing this needs assessment, you may wish to consult such resources as planning documents, demographic or corporate data, maps and training records (this assessment form also appears as Job Aid 1 in Appendix A).

1. Hazards

List the various hazards in your community or organization. What risks are you most likely to face? You can use the following check list as a starting point. **Note:** if your community has already conducted a hazard analysis, that is the best resource.
Have you completed a pandemic influenza impact assessment? Use CDC Fluaid and Flusurge as a starting point.

- ☐ Airplane crash
- ☐ Dam failure
- ☐ Drought
- ☐ Epidemic (pandemic)
- ☐ Earthquake
- ☐ Fire/firestorm
- ☐ Flood
- ☐ Hazardous material spill/release
- ☐ Hostage/shooting
- ☐ Hurricane
- ☐ Landslide/mudslide
- ☐ Mass fatality incident
- ☐ Radiological release
- ☐ Sustained power failure
- ☐ Terrorism
- ☐ Tornado
- ☐ Train derailment
- ☐ Tsunami
- ☐ Volcanic eruption
- ☐ Wildfire
- ☐ Winter storm
- ☐ Workplace violence
- ☐ Other ____________________
- ☐ Other ____________________
- ☐ Other ____________________
- ☐ Other ____________________

2. Secondary hazards

What secondary effects from those hazards are likely to impact your organization?

- ☐ Communication system breakdown
- ☐ Power outages
- ☐ Transportation blockages
- ☐ Business interruptions
- ☐ Mass evacuations/displaced population
- ☐ Overwhelmed medical/mortuary services
- ☐ Other ________________________
- ☐ Other ________________________
- ☐ Other ________________________
- ☐ Other ________________________
- ☐ Other ________________________

3. Hazard priority

What are the highest priority hazards? Consider such factors as:

- frequency of occurrence;
- relative likelihood of occurrence;
- magnitude and intensity;
- Location (affecting critical areas or infrastructure).
- spatial extent;
- speed of onset and availability of warning;
- potential severity of consequences to people, critical facilities, community functions, and property;
- Potential cascading events (e.g. damage to chemical processing plant, dam failure).

#1 Priority hazard:

#2 Priority hazard:

#3 Priority hazard:

4. Area

What geographic area(s) or facility location(s) is (are) most vulnerable to the high priority hazards?

5. Plans and procedures

What plans and procedures—emergency response plan, contingency plan, operational plan, standard operating procedures (SOPs)— will guide your organization's response to an emergency?

6. Functions

What emergency management functions are most in need of rehearsal? For example, what functions have not been exercised recently? Where have difficulties occurred in the past? You can use the following checklist as a starting point.

- ☐ Alert notification (emergency response)
- ☐ Warning (public)
- ☐ Communications
- ☐ Coordination and control
- ☐ Emergency public information (EPI)
- ☐ Damage assessment
- ☐ Health and medical
- ☐ Individual/family assistance
- ☐ Public safety
- ☐ Public works/engineering
- ☐ Transportation
- ☐ Resource management
- ☐ Continuity of government or operations
- ☐ Other ____________________________
- ☐ Other ____________________________
- ☐ Other ____________________________

7. Participants

Which agencies, departments, operational units and personnel need to participate in an exercise?

- Have any entities updated their plans and procedures?
- Have any changed policies or staff?
- Who is designated for emergency management responsibility in your plans and procedures?
- With whom does your organization need to coordinate in an emergency?
- What do your regulatory requirements call for?
- What personnel can you reasonably expect to devote to developing an exercise?

8. Programme areas

Mark the status of your emergency programme in these and other areas to identify those most in need of exercising.

	New	Updated	Exercised	Used in Emergency	N/A
Emergency plan					
Plan annex(es)					
Standard operating procedures					
Resource list					
Maps, displays					
Reporting requirements					
Notification procedures					
Mutual aid pacts					
Policy-making officials					
Coordinating personnel					
Operations staff					
Volunteer organizations					
EOC/command centre					
Communication facility					

	New	Updated	Exercised	Used in Emergency	N/A
Warning systems					
Utility emergency preparedness					
Industrial emergency preparedness					
Damage assessment techniques					
Other:					

9. Past exercises

If your organization has previously participated in exercises, what did you learn from them and what do the results indicate about future exercise needs? For example, consider the following questions:

- Who participated in the exercise, and who did not?
- To what extent were the exercise objectives achieved?
- What lessons were learnt?
- What problems were revealed, and what is needed to resolve them?
- What improvements were made following past exercises, and have they been tested?

Summary and transition

This unit provided an overview of the benefits organizations derive from exercising. Unit 2 discusses the elements of a comprehensive exercise programme and the progressive activities that make up a comprehensive programme.

Reference Library

For more information:

- WHO Exercise Development Guide for Validating Influenza Pandemic Preparedness Plans
http://www.wpro.who.int/NR/rdonlyres/DA340E3E-D27E-47A6-9833-452E7AAC9ED5/0/EDTedDRAFT1ExerciseDevelopmentGuide.pdf

- FEMA Preparedness, Training, and Exercises, Exercise FAQs:
http://www.fema.gov/

Check	**Knowledge check**

Carefully read each question and all of the possible answers before selecting the most appropriate response for each test item. Circle the letter corresponding to the answer you have chosen.

1. Research has shown that:

   a. Exercises are the best way to teach employees new skills.
   b. Extended lectures are an effective alternative to exercises.
   c. When possible, it is more effective to use real (rather than simulated) emergencies to test response procedures.
   d. People usually respond to an emergency in the way that they have trained.

2. Through exercising, organizations can reveal planning weaknesses and gaps in resources.

   a. True
   b. False

3. A good reason to exercise is to reduce the need for organizational coordination and communication.

   a. True
   b. False

4. The focus of an exercise should be on eliminating problems before an actual emergency occurs.

   a. True
   b. False

5. An exercise has value only when it leads to improvement.

   a. True
   b. False

6. In planning exercises, the main emphasis should be on the organization's:

   a. Vulnerability to natural hazards.
   b. Vulnerability to man-made hazards (e.g. terrorism).
   c. Ability to effectively carry out response and recovery functions.
   Ability to respond to specific types of emergencies.

Answers	**Knowledge Check (Continued)**
	1. d 2. a 3. b 4. a 5. a 6. c

Unit 2: Comprehensive Exercise Programme

UNIT 2

COMPREHENSIVE EXERCISE PROGRAMME

Introduction

In any discussion of emergency preparedness, for any hazard and most especially in the context of planning for a pandemic, the emphasis must be on a comprehensive exercise programme, made up of progressively complex exercises, each one building on the previous one, until the exercises are as close to reality as possible. This unit provides an overview of five main types of exercise activities that make up a comprehensive exercise programme.

Unit 2 objectives

After completing this unit, you should be able to:

- identify the basic components of a comprehensive exercise programme;
- explain the importance of designing a comprehensive and progressive exercise programme to meet the needs of your organization, community or nation.

Progressive exercising

A progressive programme has several important characteristics.

- The exercise programme involves the efforts and participation of various entities, such as departments, organizations, or agencies. Through the involvement of multiple entities, the programme allows the involved organizations to not only test their implementation of emergency management procedures, but also their coordination with each other in the process.

- The programme is carefully planned to achieve identified goals.

- The programme is made up of a series of increasingly complex exercises.

- Each successive exercise builds upon the previous one until mastery is achieved.

Progressive Exercising (Continued)

Broad commitment

A community or organization engaging in a progressive exercise programme must ensure it is comprehensive and takes into account all responding agencies and organizations.

Communities are composed of more than public health agencies, police and fire authorities, and public works. The following entities also have requirements for exercising.

- Hospitals.
- Airports, chemical and nuclear facilities, and other regulated organizations.
- Volunteer agencies and organizations in the private sector that contribute services, materials and personnel to response and recovery efforts.
- Public or private work sites that may be vulnerable to significant emergency events.

A progressive exercise programme, therefore, requires a commitment from various agencies and organizations to participate in increasingly challenging exercises in order to address the larger emergency management system rather than a single problem.

The same is true when a single organization engages in a progressive exercise programme. The role of each department and function involved in responding to and recovering from an emergency event must be considered, and the organization must secure the commitment of all of these elements to a sequence of progressive internal and external exercises, with the goal of building a coordinated, effective response.

Careful planning

Exercises require careful planning around clearly identified goals. Only through identifying exercise goals, then designing, developing, conducting and analyzing the results, can those responsible for emergency operations be sure of what works and what does not.

Progressive Exercising (Continued)

Increasing complexity

Exercises should be organized to increase in complexity—for example, from tabletop discussions to functional exercises to a full-scale exercise. Each type of exercise builds on previous exercises using more sophisticated simulation techniques and requiring more preparation time, personnel, and planning.

Rushing into a full-scale exercise too quickly can open the door to potential failure because shortfalls have not been identified through less complicated and expensive exercises.

Success breeds success

An important advantage of building incrementally to a full-scale exercise is that successful exercise experiences breed new successes.

- Officials and stakeholders are more willing to commit resources.
- Personnel are more motivated and look forward to the next exercise.
- Confidence increases.
- Operating skills improve.

Who participates?

For a local or nationwide exercise programmes, the jurisdiction (usually the Ministry of Health for pandemic exercises) determines what agencies, organizations, and stakeholders participate in each exercise. Participants are further determined by the nature and size of the exercise. Larger exercises would include all of the participants who would have responsibilities in a real pandemic or other emergency. Smaller exercises, which focus on a limited aspect of the pandemic response emergency plan, would limit the participants.

Who Participates? (Continued)

The same is true of exercises conducted by any particular organization. For example, let's look at the case of a vaccine manufacturer which, during an influenza pandemic, converts its production line over to influenza vaccine production for mass distribution. This organization might design exercises to test procedures for the following.

- Coordination with jurisdiction officials.
- Managerial decision-making on when to convert, how much of the line to convert, and when to convert back.
- Internal notifications.
- Line personnel responsibilities.
- Temporary facility changes.
- Coordination with suppliers.
- Product distribution.
- Transitioning back to regular production.
- Documentation.

In either case, whether the exercises involve an entire community or a more limited population or sector, the nature of the exercise somewhat determines the participants. On one end of the spectrum, a tabletop exercise might involve only key decision-makers. An exercise to test particular functions would limit its participants to those functions, and a full-scale exercise might involve the entire facility, ministry, community or nation.

Some types of exercises have additional participant requirements. For example, a functional exercise involves not only players, but also simulators, controllers and evaluators.

What activities are included?

There are five main types of activities in a comprehensive exercise programme:

- orientation seminar(s)
- drill(s)
- tabletop exercises
- functional exercises
- a full-scale exercise.

As previously discussed, these activities build from simple to complex, from narrow to broad, from least expensive to most costly to implement, from theoretical to realistic. When carefully planned to achieve specified objectives and goals, this progression of exercise activities provides an important element of an integrated emergency preparedness system.

Orientation seminar

Purpose

As the name suggests, the **orientation seminar** is an overview or introduction. Its purpose is to familiarize participants with roles, plans, procedures or equipment. It can also be used to resolve questions of coordination and assignment of responsibilities.

Characteristics

Key characteristics of the orientation seminar are summarized in the table on the next page.

What Activities Are Included? (Continued)

Orientation seminar characteristics	
Format	The orientation seminar is a very low-stress event, usually presented as an informal discussion in a group setting. There is little or no simulation and for this reason, orientations are not usually recognized exercises, but are an important part of the cycle. A variety of seminar formats can be used, including: ▪ lecture ▪ discussion ▪ slide or video presentation ▪ computer demonstration ▪ panel discussion ▪ guest lecturers.
Applications	The orientation seminar can be used for a wide variety of purposes. ▪ Discussing a topic or problem in a group setting. ▪ Introducing something new (e.g. policies, plans and resources). ▪ Explaining existing plans to new people (e.g. staff, newly-elected officials or executives need an explanation of the EOP and their role at the EOC and new employees need an orientation to operational plans as they relate to emergencies). ▪ Introducing a cycle of exercises or preparing participants for success in more complex exercises. ▪ Motivating people for participation in subsequent exercises.
Leadership	Orientations are led by a facilitator, who presents information and guides discussion. The facilitator should have some leadership skills, but very little other training is required.
Participants	A seminar may be cross-functional, involving one or two participants from each function or service being discussed (e.g. management, policy, coordination and operations staff), or it may be geared to several people from a single agency or department.
Facilities	A conference room or any other fixed facility may be used, depending on the purposes of the orientation.
Time	Orientations should last a maximum of 1–2 hours.
Preparation	An orientation is quite simple to prepare and conduct, with a preparation time of two weeks usually sufficient. Participants need no previous training.

What Activities Are Included? (Continued)

Conducting an orientation

There are no cut-and-dried rules for an effective orientation; its purpose will determine its format. Below are a few helpful guidelines for conducting a seminar.

Orientation seminar guidelines

- **Be creative.** You can use various discussion and presentation methods. Think of interesting classes that you have attended in other subjects, and borrow the techniques of good teachers and presenters. For example, you might call on people one by one to give ideas, plan a panel discussion, hold a brainstorming session, present case studies for problem-solving, or give an illustrated lecture.
- **Get organized and plan ahead.** Even though orientation seminars are less complex than other activities, it is no time to "wing it."
- **Be ready to facilitate a successful orientation seminar.** Discourage long discussions, keep exchanges crisp and to the point, focus on the subject at hand and help everyone feel good about being there.

Drills

Purpose

A drill is a coordinated, supervised exercise activity, normally used to test a single specific operation or function. With a drill, there is no attempt to coordinate organizations or fully activate the EOC. Its role in an exercise programme is to practice and perfect one small part of the response plan and help prepare for more extensive exercises, in which several functions will be coordinated and tested. The effectiveness of a drill is its focus on a single, relatively limited portion of the overall emergency management system. It makes possible a tight focus on a potential problem area.

Characteristics

Key characteristics of drills are summarized in the table on the next page.

What Activities Are Included? (Continued)

Drill characteristics	
Format	A drill involves actual field or facility response for an EOC operation. It should be as realistic as possible, employing any equipment or apparatus for the function being drilled.
Applications	Drills are used to test a specific operation. They are also used to provide training with new equipment, to develop new policies or procedures, or to practice and maintain current skills. Drills are a routine part of the daily job and organizational training in the field, in a facility, or at the EOC. Some examples of drills run by different organizations are listed below. ▪ EOC: call down procedures. ▪ Staff calls back and telephone trees. ▪ Public health and safety: site assessment and sampling. ▪ Red Cross: locating specific types of blood within a time constraint. ▪ Military: activation and mobilization drill. ▪ Airport: fire department response to the furthest part of a runway within a given time. ▪ Chemical plant: evacuation and isolation of spill area and valve system shutoff ▪ Private sector resource provider: warehouse readiness drill.
Leadership	A drill can be led by a manager, supervisor, department head, or exercise designer. Staff must have a good understanding of the single function being tested.
Participants	The number of participants depends on the function being tested. Coordination, operations, and response personnel could be included.
Facilities	Drills can be conducted within a facility, in the field, or at the EOC or other operating centre.
Time	Between 30 minutes to 2 hours is usually required.
Preparation	Drills are one of the easiest kinds of exercise activities to design. Preparation may take about a month. Participants usually need a short orientation beforehand.

What Activities Are Included? (Continued)

Conducting a drill

How a drill is conducted varies according to the type of drill, ranging from simple operational procedures to more elaborate communication and command post drills. For example, a command post drill would require participants to report to the drill site, where a "visual narrative" would be displayed in the form of a mock emergency. Equipment, such as vans, command boards, and other needed supplies would be made available.

Given the variety of functions that may be drilled, there is no set way to run a drill. However, some general guidelines are given below.

Drill guidelines

- **Prepare.** If operational procedures are to be tested, review them beforehand. Review safety precautions.
- **Set the stage.** It's always good to begin with a general briefing, which sets the scene and reviews the drill purpose and objectives. Some designers like to set the scene using films, slides, or videotapes.
- **Monitor the action.** After a drill has been started, it will usually continue under its own steam. If you find that something you wanted to happen is not happening, however, you might want to insert a message to trigger that action (you will learn more about messages later on).

What Activities Are Included? (Continued)

Tabletop exercise

Purpose

A tabletop exercise is a facilitated analysis of an emergency situation in an informal, stress-free environment. It is designed to elicit constructive discussion as participants examine and resolve problems based on existing operational plans and identify where those plans need to be refined. The success of the exercise is largely determined by group participation in the identification of problem areas.

Characteristics

There is minimal attempt at simulation in a tabletop exercise. Equipment is not used, resources are not deployed, and time pressures are not introduced. Key characteristics of the tabletop exercise are summarized in the table on the next page. We will discuss how to conduct a tabletop exercise in Unit 5.

What Activities Are Included? (Continued)

<table>
<tr><th colspan="2">Tableteop exercise characteristics</th></tr>
<tr><td>Format</td><td>The exercise begins with the reading of a short narrative, which sets the stage for the hypothetical emergency. Then, the facilitator may stimulate discussion in two ways.

▪ Problem statements. Problem statements (describing major or detailed events) may be addressed either to individual participants or to participating departments or agencies. Recipients of problem statements then discuss the actions they might take in response.
▪ Simulated messages. These messages are more specific than problem statements. Again, the recipients discuss their responses.

In either case, the discussion generated by the problem focuses on roles (how the participants would respond in a real emergency), plans, coordination, the effect of decisions on other organizations, and similar concerns. Often maps, charts, and packets of materials are used to add to the realism of the exercise.</td></tr>
<tr><td>Applications</td><td>Tabletop exercises have several important applications.

▪ They lend themselves to low-stress discussion of coordination and policy.
▪ Provide a good environment for problem-solving.
▪ Provide an opportunity for key agencies and stakeholders to become acquainted with one another, their interrelated roles, and their respective responsibilities.
▪ Provide good preparation for a functional exercise.</td></tr>
<tr><td>Leadership</td><td>A facilitator leads the tabletop discussion. This person decides who gets a message or problem statement, calls on others to participate, asks questions, and guides the participants toward sound decisions.</td></tr>
<tr><td>Participants</td><td>The objectives of the exercise dictate who should participate. The exercise can involve many people and many organizations—essentially anyone who can learn from or contribute to the planned discussion items. This may include all entities that have a policy, planning, or response role.</td></tr>
<tr><td>Facilities</td><td>A tabletop exercise requires a conference or meeting room where participants can gather around a table.</td></tr>
<tr><td>Time</td><td>A tabletop exercise usually lasts from 1 to 4 hours but can be longer. Discussion times are open-ended, and participants are encouraged to take their time in arriving at in-depth decisions without time pressure. When the time is up, the activity is concluded. Although the facilitator maintains an awareness of time allocation for each area of discussion, the group does not have to complete every item in order for the exercise to be a success.</td></tr>
<tr><td>Preparation</td><td>It typically takes about a month to prepare for a tabletop exercise. Preparation also usually requires at least one orientation and sometimes one or more drills.</td></tr>
</table>

What Activities Are Included? (Continued)

Functional exercises

Purpose

A functional exercise is a fully simulated interactive exercise that tests the capability of an organization to respond to a simulated event. The exercise tests multiple functions of the organization's operational plan. It is a coordinated response to a situation in a time-pressured, realistic simulation.

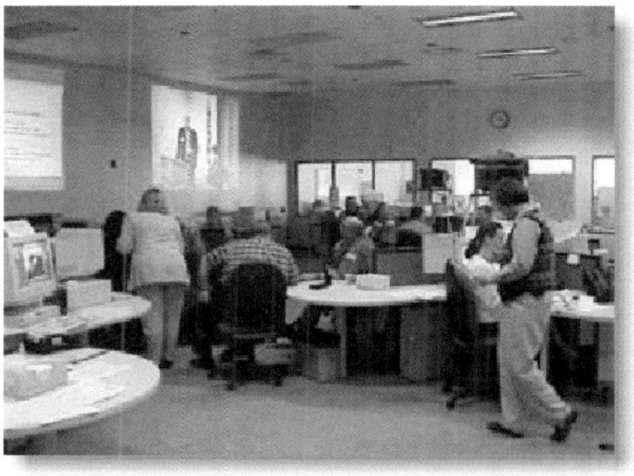

Characteristics

A functional exercise focuses on the coordination, integration, and interaction of an organization's policies, procedures, roles and responsibilities before, during, or after the simulated event. Key characteristics of a functional exercise are summarized in the following table. We will discuss how to conduct a functional exercise in Unit 6.

What Activities Are Included? (Continued)

<table>
<tr><th colspan="2">Functional exercise characteristics</th></tr>
<tr><td>Format</td><td>This is an interactive exercise, similar to a full-scale exercise without the equipment. It simulates an incident in the most realistic manner possible short of moving resources to an actual site. A functional exercise incorporates the following.

▪ Geared for policy, coordination and operations personnel. The "players" in the exercise practice responding in a realistic way to carefully planned and sequenced messages given to them by "simulators." The messages reflect ongoing events and problems that might actually occur in a real emergency.

▪ A stressful exercise, because players respond in real time, with on-the-spot decisions and actions. All of the participants' decisions and actions generate real responses and consequences from other players.

▪ Complex—messages must be carefully scripted to force participants to make decisions and act on them. This complexity makes the functional exercise difficult to design.</td></tr>
<tr><td>Applications</td><td>Functional exercises make it possible to test several functions and exercise several agencies or departments without incurring the cost of a full-scale exercise. A functional exercise is always a prerequisite to a full-scale exercise.

In some instances, taking part in a functional exercise may serve as a full-scale exercise for a participating organization (e.g. a hospital may conduct its own full-scale exercise as part of a community-based functional exercise).</td></tr>
<tr><td>Leadership and participants</td><td>Functional exercises are complex in their organization of leadership and the assignment of roles. The following general roles are used.

▪ Controller: manages and directs the exercise.
▪ Players: participants who respond as they would in a real emergency (players should include policy-makers and may include coordinators and operational personnel directing field activities).
▪ Simulators: assume external roles and deliver planned messages to the players
▪ Evaluators: observers who assess performance.</td></tr>
<tr><td>Facilities</td><td>It is usually conducted in the EOC or other operating centre. Ideally, people gather where they would actually operate in an emergency. Players and simulators are often seated in separate areas or rooms. Realism is achieved by the use of telephones, radios, televisions and maps.</td></tr>
</table>

What Activities Are Included? (Continued)

Functional Exercise Characteristics (Continued)	
Time	A functional exercise requires from 3 to 8 hours, although it can run a full day or even longer.
Preparation	Plan on 6 to 18 months or more to prepare for a functional exercise, for several reasons: ▪ Staff members need considerable experience with the functions being tested. ▪ The exercise should be preceded by lower-level exercises, as needed. ▪ The controller, evaluators, and simulators require training. ▪ The exercise may require a significant allocation of resources and a major commitment from organizational leaders.

What Activities Are Included? (Continued)

Full-scale exercise

Purpose

A **full-scale exercise** simulates a real event as closely as possible and is designed to evaluate the operational capability of emergency management systems in a highly stressful environment simulating actual response conditions. To accomplish this realism, requires the mobilization and movement of emergency personnel, equipment and resources. Ideally, the full-scale exercise should test and evaluate most functions of the emergency management plan or operational plan.

Characteristics

A full-scale exercise differs from a drill in that it coordinates the actions of several entities, tests several emergency functions, and activates the EOC or other operating centre. Realism is achieved through:

- on-scene actions and decisions from policy groups
- Simulated "victims"
- rapid detection, reporting and response requirements
- communication devices
- equipment deployment
- actual resource and personnel allocation.

Key characteristics of full-scale exercises are summarized in the table on the next page. We will discuss how to conduct a full-scale exercise in Unit 7.

What Activities Are Included? (Continued)

Full-scale exercise characteristics	
Format	The exercise begins with a description of the event, communicated to responders in the same manner as would occur in a real event. Personnel conducting the field component must proceed to their assigned locations, where they see a "visual narrative" in the form of a mock emergency (e.g. a plane crash with victims, a "burning" building, a simulated chemical spill on a highway, or a terrorist attack). From then on, actions taken at the scene serve as input to the simulation taking place at the EOC or operating centre.
Applications	Full-scale exercises are the ultimate in the testing of functions—the "trial by fire." Because they are expensive and time consuming, it is important that they be reserved for the highest priority hazards and functions.
Leadership and participants	One or more controllers manage the exercise, and evaluators are required. All levels of personnel take part in a full-scale exercise: ▪ policy personnel ▪ coordination personnel ▪ operations personnel ▪ field personnel.
Facilities	The event unfolds in a realistic setting (e.g. outbreak in a community, an IHR event attack at a public venue). The EOC or other operating centre is activated, and field command posts may be established.
Time	A full-scale exercise may be designed to be as short as 2 to 4 hours, or to last as long as one or more days.
Preparation	Preparation for a full-scale exercise requires an extensive investment of time, effort and resources—1 to 1½ years to develop a complete exercise package. This time frame includes multiple drills and preparatory tabletop and functional exercises. In addition, personnel and equipment from participating agencies must be committed for a prolonged period of time.

Comparing the five activities

Each of the five activities just described play an important part in the overall exercise programme. The following chart lists some of reasons for conducting each type of activity. Key characteristics of each type of exercise are shown in the table on the next page.

Reasons to conduct exercise programme activities				
Orientation	**Drill**	**Tabletop exercise**	**Functional exercise**	**Full-scale exercise**
No previous exercise.	Assess equipment capabilities	Practice group problem solving	Evaluate a function	Assess and improve information analysis
No recent operations	Test response time	Promote executive familiarity with emergency management plan	Observe physical facilities use	Assess and improve interagency cooperation
New plan	Personnel training	Assess plan coverage for a specific case study	Reinforce established policies and procedures	Support policy formulation
New procedures	Assess interagency cooperation	Assess plan coverage for a specific risk area	Assess hospital preparedness	Assess negotiation procedures
New staff, leadership	Verify resource and staffing capabilities	Examine staffing contingencies	Test seldom-used resources	Test resource and personnel allocation
New biological risk		Assess interagency or interdepartmental coordination	Assess and strengthen inter-jurisdictional or inter-organizational relations	Assess and strengthen inter-jurisdictional or inter-organizational relations
		Observe information sharing		Assess personnel and equipment locations
		Train personnel in negotiation		Test equipment capabilities

Comparing the Five Activities (Continued)

Key characteristics

The following table briefly compares the key characteristics of the five types of exercise programme activities.

Comparison of key activity characteristics					
	Orientation	**Drill**	**Tabletop exercise**	**Functional exercise**	**Full-scale exercise**
Format	Informal discussion in group setting Various presentation methods	Actual field or facility response Actual equipment	Narrative presentation Problem statements or simulated messages Group discussion No time pressures	Interactive, complex Players respond to messages (events/problems) provided by simulators Realistic but no actual equipment Conducted in real time; stressful	Realistic event announcement Personnel gather at assigned site Visual narrative (enactment) Actions at scene serve as input to EOC simulation
Leaders	Facilitator	Manager, supervisor, department head, or designer	Facilitator	Controller	Controller(s)
Participants	Single agency/ department, or cross-functional	Personnel for the function being tested May include coordination, operations, response personnel	Anyone with a policy, planning, or response role for the type of situation used	Players (policy, coordination, and operations personnel) Simulators Evaluators	All levels of personnel (policy, coordination, operations, field) Evaluators
Facilities	Conference room	Facility, field, or EOC	Large conference room	EOC or other operating centre (multiple rooms)	Realistic setting EOC or other operating centre
Time	1–2 hours	½–2 hours	1–4 hours or longer	3–8 hours or longer	2 hours to 1 or more days

Comparing the Five Activities (Continued)

Comparison of Key Activity Characteristics (Continued)					
	Orientation	**Drill**	**Tabletop Exercise**	**Functional Exercise**	**Full-scale Exercise**
Preparation	Simple preparation, two weeks	Easy to design, one month Participants need orientation	One month preparation Preceded by orientation and 1 or more drills	Complex, 6–18 months preparation Preceded by simpler exercises Significant allocation of resources	Extensive time, effort, resources One year to 18 months development Includes preparatory drills, tabletops, functional exercises

Building an exercise programme

As you have seen, a progressive exercise programme involves the combined efforts of many agencies, departments, or other entities in a series of activities that increase in complexity until mastery is achieved.

Building an exercise programme is a little like planning a single exercise—except that the activities take place on a much larger scale. Plans are developed by a team and are based on a careful examination of the operating plan.

The development of an exercise programme has many facets, including:

- analysis of capabilities and costs
- scheduling of tasks
- public relations efforts
- development of a long-term plan.

Careful work on the long-term plan will carry over into the design of individual exercises.

The planning team

Building an Exercise Programme (Continued)

In a private or volunteer organization that is planning an exercise programme, the team would be similarly organized, with representatives of all major functions and departments.

The emergency manager and other emergency personnel (or the person responsible for the organization's emergency response effort) would take the lead, and the representatives would then meet to analyze what they need to do to support one another. Often organizations can meet the exercise needs of more than one agency at a time. This teamwork can help establish important relationships among participating organizations.

Later, members of the team can also serve on exercise design teams to design individual exercises.

Goal setting

Because a comprehensive exercise programme usually extends over several months, it is important to set long-term goals or develop a mission statement. Without this, the programme is likely to lack focus and continuity.

Schedule and sequence

When these preliminary steps (organizing the team and establishing a mission statement and goals) have been taken, the hard work of drawing up a plan can take place. Developing the exercise programme plan involves:

1. laying out a series of exercises that can meet the needs of the various participating entities;
2. organizing them into a workable sequence and time schedule.

Building an Exercise Programme (Continued)

Plan format

An exercise programme plan can use any format, but it should include the following elements:

- a time frame
- a problem statement
- long-range goal(s)
- functional objectives
- a schedule.
- Exercise descriptions, including:
    - type of exercise
    - participants
    - purpose
    - rationale.

A sample plan format is shown on the next page. This is a hypothetical example of one community's exercise plan.

Building an Exercise Programme (Continued)

Plan Format	Sample plan: Comprehensive Exercise Programme
	Note: during the previous year, several tabletops and functional exercises were held to test weaknesses in Communications, Alert and Notification, and Individual/Family Assistance during communicable disease outbreaks. The series of exercises might take less time in some communities.
Time frame	The exercise programme extends over an 18-month period.
Present problems	This programme has been formulated to address problems arising as a result of rapid population growth and growing anxiety versus the possibility of a pandemic strain emerging. According to experts, possibilities for a pandemic are increasing. Personnel involved in the functional areas listed below have not been tested in the last year.
Long-range goal	To work toward a full-scale exercise testing all important functions in the context of a mass casualty incident. This will satisfy regulatory requirements and full-scale exercise requirements for the hospital and airport, by involving these agencies.
Functions to be tested	Health and Medical, Public Information, Coordination and Control (EOC Operations, Incident Command). ▪ To determine the adequacy of plans and procedures within the following functional areas to handle a mass casualty incident: Health and Medical, Public Information, Coordination and Control (EOC Operations, Incident command). ▪ To test the ability of the above-named functional areas to communicate and coordinate their response efforts during a potential pandemic incident. ▪ To test the ability to obtain adequate resources (locally and through mutual aid agreements) in the above-named functional areas to handle a large communicable disease incident.
First month	Exercise: Orientation For: Emergency management, health departments, staff and heads of various agencies, such as a mental health association, funeral director, coroner, fire and police. Purpose: To review new plans and procedures for dealing with mass casualty incidents. Rationale: Inform those who are unaware of plans and gain support and additional input from department leaders.

Building an Exercise Programme (Continued)

Plan format	Sample plan: Comprehensive Exercise Programme	
Second month	Exercise:	Orientation
	For:	Emergency management, health department heads, staff and heads of various agencies: fire, police staff, county public information officer (PIO).
	Purpose:	To review new plans for large communicable disease incidents with responders.
	Rationale:	Gain support and additional input from first responders and acquaint them with leadership's plans.
Fourth month	Exercise:	Training course with functional exercise.
	For:	Responders and incident commanders; emergency management staff; health staff, various chiefs, captains, lieutenants from fire and police; emergency medical services (EMS), mental health, Radio Amateur Civil Emergency Services (RACES), funeral directors, county coroner, county PIO.
	Purpose:	To provide training in incident response.
	Rationale:	This is a training session in CD Incident Response Course This course provides an excellent overview of specific needs related to CD incidents. The course culminates in a functional activity.
Seventh month	Exercise:	Drill
	For:	Fire, police, EMS, coroner, funeral directors.
	Purpose:	To set up the Incident Command System (ICS) for responding CD incidents.
	Rationale:	Establish ICS to support needed functions and tasks.
Eighth month	Exercise:	Drill.
	For:	PIO, Health, police, Emergency Management.
	Purpose:	To set up a Joint Information Centre (JIC).
	Rationale:	Acquaint participants with the PIO function and JIC operations, test equipment and lines of communication.

Building an Exercise Programme (Continued)

Plan Format	Sample plan: Comprehensive Exercise Programme	
Ninth month	Exercise:	Drill.
	For:	Mental health, funeral directors, PIO, clergy, Emergency Manager.
	Purpose:	To set up a family assistance centre.
	Rationale:	Acquaint participants with the office equipment and test role as support to the victims' families.
Eleventh month	Exercise:	Tabletop exercise.
	For:	Incident Command, PIO, health services, police, fire and EMS.
	Purpose:	To pull together the three functions tested in the previous drills in the context of a CD incident as the result of a outbreak.
	Rationale:	Address and resolve potential communication and coordination problems among the Incident Command, PIO, health services, police, fire, and EMS, before the functional exercise.
Fourteenth month	Exercise:	Functional exercise.
	For:	Communications, coordination and control, ICS and EOC, PIO, health and medical.
	Purpose:	To test additional functions for mass fatality in the context of a CD Event: emergency public information effectiveness, health and medical mass casualty, coordination and control, ICS, and EOC operations.
	Rationale:	Identify preliminary shortfalls and test overall coordination before full-scale exercise.
Fifteenth month	Exercise:	Tabletop exercise.
	For:	Communications, coordination and control, ICS and EOC, PIO, health and medical.
	Purpose:	To correct and retest problems identified in preceding functional exercise.
	Rationale:	Work out potential problems discovered in the previous functional exercise and make adjustments necessary before the full-scale exercise.

Building an Exercise Programme (Continued)

Plan Format	Sample plan: Comprehensive Exercise Programme	
Eighteenth month	Exercise:	Full-scale exercise: pandemic influenza.
	For:	All agencies: heads of agencies and responders.
	Purpose:	To test all functions in the context of a pandemic influenza.
	Rationale:	The exercise fulfils full-scale requirements for plan validation requirements for managing the consequences of a pandemic and on accreditation of health care organizations certification for the hospital.

Activity

Activity: develop a Comprehensive Exercise Programme Plan

Working from the needs assessment and specifically the pandemic impact assessment you completed for your jurisdiction or organization in the previous unit, develop a plan for a comprehensive exercise programme to address those needs. Include the key elements discussed in the last section. You can use the following worksheet (which also appears as Job Aid 2 in Appendix A) in developing your plan. If this format doesn't work for you, change it to meet your needs.

Comprehensive Exercise Programme Planning Worksheet

Time frame:

Present problems:

Long-range goal:

Functional objectives:

Month:

Exercise:

For:

Purpose:

Rationale:

Activity: Develop a Comprehensive Exercise Programme Plan (Continued)

Month:

Exercise:

For:

Purpose:

Rationale:

Month:

Exercise:

For:

Purpose:

Rationale:

Month:

Exercise:

For:

Purpose:

Rationale:

Month:

Exercise:

For:

Purpose:

Rationale:

Activity: Develop a Comprehensive Exercise Programme Plan (Continued)

Month:

Exercise:

For:

Purpose:

Rationale:

Month:

Exercise:

For:

Purpose:

Rationale:

Month:

Exercise:

For:

Purpose:

Rationale:

Month:

Exercise:

For:

Purpose:

Rationale:

Activity: Develop a Comprehensive Exercise Programme Plan (Continued)

Month:

Exercise:

For:

Purpose:

Rationale:

Month:

Exercise:

For:

Purpose:

Rationale:

Summary and transition

Unit 2 provided an overview of the five main types of exercise activities that make up a comprehensive exercise programme. Unit 3 provides general information on the exercise development process and illustrates how the activities in that process relate to one another.

Reference Library

Check

Knowledge check

Carefully read each question and all of the possible answers before selecting the most appropriate response for each test item. Circle the letter corresponding to the answer you have chosen.

1. In a progressive exercise programme, the exercises:

   a. Have a consistent format but are conducted with increasing frequency.
   b. Are organized to increase in complexity.
   c. Are organized to decrease in complexity.
   d. Are sponsored on a rotating basis by different organizations.

2. A likely cause of exercise failure is:

   a. Running too many drills and functional exercises.
   b. Conducting orientations and drills before functional exercises.
   c. Rushing into a full-scale exercise before the organization is ready.
   d. Basing the selection of participants on the nature of the exercise.

3. Which statement is true of an orientation?

   a. It requires field sites and actual equipment.
   b. It may be used to introduce or explain plans and policies.
   c. It involves a controller, simulators, and evaluators.
   d. It is used to test a specific operation.

4. Which statement is true of a drill?

   a. It is best conducted in a conference room.
   b. It involves a controller, simulators and evaluators.
   c. It is used to test a specific operation.
   d. It is aimed primarily at policy-makers and decision-makers.

5. Which statement is true of a tabletop exercise?

   a. It involves a highly realistic simulation.
   b. It involves a controller, simulators and evaluators.
   c. It requires field sites and actual equipment deployment.
   d. It is a facilitated analysis of an emergency situation.

6. Which statement is true of a functional exercise?

   a. It involves a controller, simulators and evaluators.
   b. It is simple, informal, and stress-free.
   c. It requires field sites and actual equipment deployment.
   d. It may be used to introduce or explain plans and policies.

Knowledge Check (Continued)

7. Which statement is true of a full-scale exercise?

   a. It involves a highly realistic simulation.
   b. It is aimed primarily at policy-makers and decision-makers.
   c. It requires field sites but actual equipment remains in the shed.
   d. It is used to test a specific operation.

8. Development of an exercise programme includes analysis of capabilities and costs and scheduling of tasks.

   a. True
   b. False

9. The planning for an exercise programme should be done primarily by the emergency manager or whoever is responsible for the organization's emergency response effort.

   a. True
   b. False

10. Which statement is **NOT** true of progressive exercise programmes?

    a. They allow participating organizations to test both implementation of procedures and coordination with each other.
    b. Each successive exercise builds upon the previous one until mastery is achieved.
    c. The entire programme is planned to achieve identified goals.
    d. Very little commitment is required from participating agencies because they can opt in or out at any time.

Answers

Knowledge Check (Continued)

1. b
2. c
3. b
4. c
5. d
6. a
7. a
8. a
9. b
10. d

Unit 3: The Exercise Process

Introduction

In the previous unit, we took the broad view of planning of an entire comprehensive exercise programme. In this unit, we will take a step closer and look at what goes into planning and implementing a single exercise within that programme.

When an exercise proceeds smoothly, it all looks so easy! But there is far more to it than the time spent in the exercise itself. A great deal of thought and planning preceded the exercise, and more work will follow.

> An exercise is not an independent activity with clearly-marked beginning and ending points. Rather, it can be seen as part of a complex process that involves a number of accomplishments and tasks. All of those tasks are interrelated and affect not only the success of the current exercise, but the design and success of future exercises.

It will be much easier to understand and visualize the individual activities if you can first see how they fit into the overall process. This unit presents an overview of the main exercise activities and their relationships to one another. It also discusses some important aspects of laying the groundwork for an exercise. In later units, we will focus more closely on particular aspects of the design process.

Unit 3 objectives

After completing this unit, you should be able to:

- identify the five major accomplishments in designing and implementing an exercise;
- describe the organization of an exercise design team;
- define the purpose of the four exercise documents.

The big picture

In preparation for launching an exercise programme and designing individual exercises, it is important to have a clear vision of the entire exercise process. There are a number of ways to visualize the exercise process. Let's look briefly at three graphic representations of the process: organized by task sequence, organized by task categories and phase, and organized by major accomplishments. Each of them helps clarify the overall process.

Sequence of main tasks

In the following graphic, the main tasks are shown in their approximate sequence. This chart may help you get a good mental picture of the entire sequence. It is also a good place to start in creating a more detailed schedule of tasks.

Sequence of tasks for a successful exercise

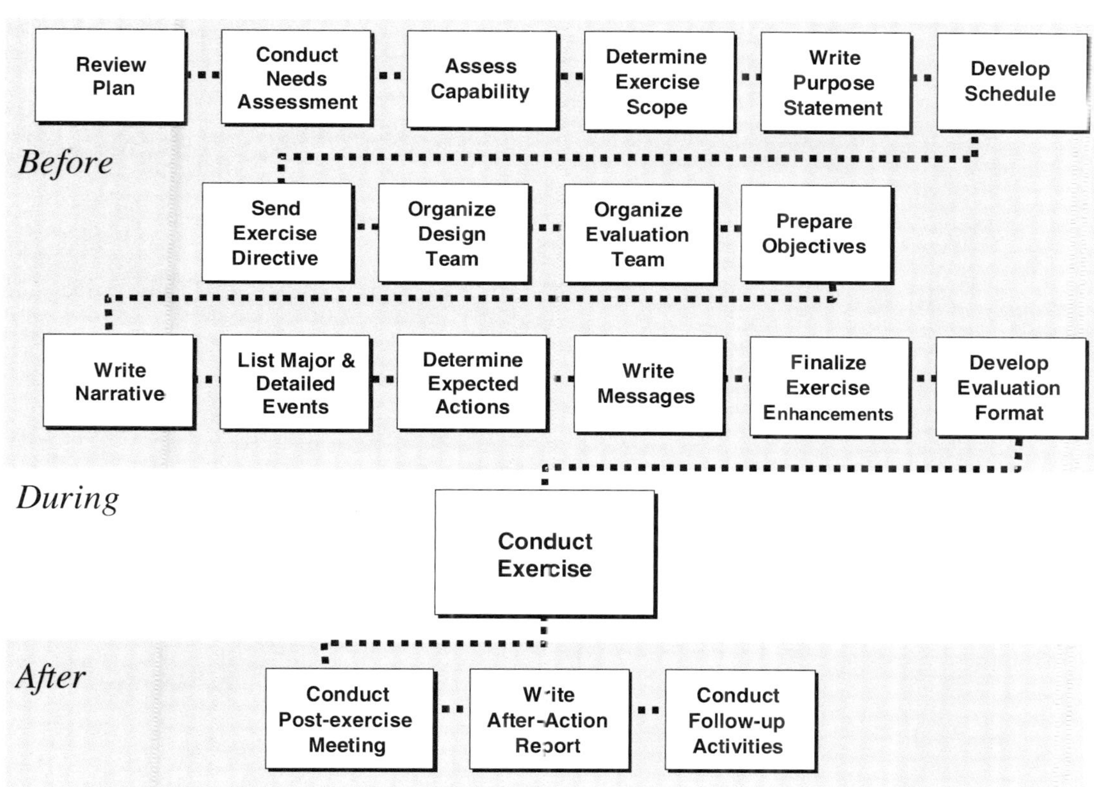

The Big Picture (Continued)

Categories of tasks

Another way to look at the exercise process is by organizing the tasks into two dimensions.

- Exercise phase (pre-exercise, exercise, and post-exercise).
- Type of task (those related to design and those related to evaluation).

The following matrix illustrates this type of organization.

Task categories			
	Pre-exercise Phase	**Exercise Phase**	**Post-exercise Phase**
Design	▪ Review plan ▪ Assess capability ▪ Address costs and liabilities ▪ Gain support/issue exercise directive ▪ Organize design team ▪ Draw up a schedule ▪ Design exercise (eight design steps)	▪ Prepare facility ▪ Assemble props and other enhancements ▪ Brief participants ▪ Conduct exercise	
Evaluation	▪ Select evaluation team leader ▪ Develop evaluation methodology ▪ Select and organize evaluation team ▪ Train evaluators	▪ Observe assigned objectives ▪ Document actions	▪ Assess achievement of objectives ▪ Participate in post exercise meetings ▪ Prepare evaluation report ▪ Participate in follow up activities

The Big Picture (Continued)

Major task accomplishments

One of the simplest ways to envision the exercise process is by major accomplishments. As shown in the graphic below, the process can be factored into five major accomplishments that make up the design cycle.

1. Establishing the base
2. Exercise development
3. Exercise conduct
4. Exercise critique and evaluation
5. Exercise follow up

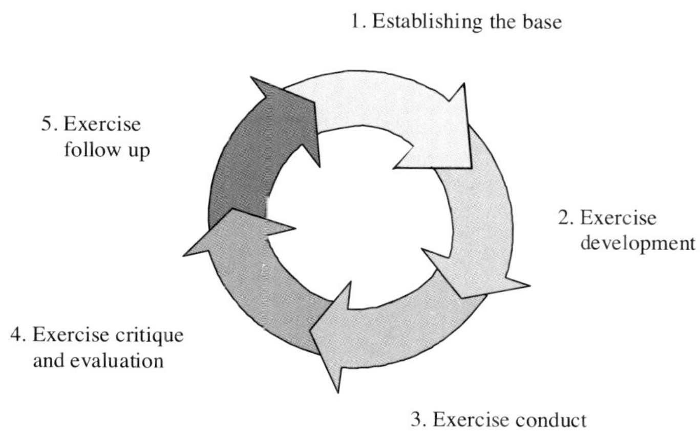

Each accomplishment is the outgrowth of a set of specific tasks and subtasks (similar to those listed in the earlier models), which we will review shortly. The process is circular, with the results of one exercise providing input for the next.

The Big Picture (Continued)

Flexibility is key

The exercise process applies no matter what level of exercise is being designed and no matter what the size of the organization using it. Whether you are located in a large jurisdiction with extensive resources, a smaller community with meagre resources, or a private or volunteer organization with a particular vested interest in emergency preparations, you can use this process.

The underlying premise is that this process must be flexible enough to meet the unique exercise needs of the organization using it. Therefore, as you consider each task, it is important to remember that each task must be designed, tailored and applied in a manner that suits **your** organization's specific objectives and capabilities.

Accomplishment 1: establishing the base

Exercises are designed to motivate personnel to think or act as they would in a real event. Establishing the base is basically laying the groundwork for the exercise to ensure that this is achieved. Getting ready for the exercise involves the following.

1. **Reviewing the current plan.** What does it tell us about ideal performance i.e. how are we supposed to implement policies and procedures in the event of an emergency?

2. **Conducting a needs assessment.** What are our risks and vulnerabilities, and where do we need to focus our training efforts?

3. **Assessing the jurisdiction's capability to conduct an exercise.** What resources can we draw from to design and implement an exercise?

4. **Defining the exercise scope.** How should we limit this exercise?

5. **Selecting the exercise type.** What type of exercise best meets our training needs within the available resources?

6. **Addressing the costs and liabilities.** What will the exercise cost in terms of funding, human resources and organizational liability?

7. **Developing a statement of purpose.** What do we expect to gain from the exercise?

Accomplishment 1: Establishing the Base (Continued)

8. **Gain support and announce the exercise.** How can we obtain the support of those in authority and then use that support to garner support among participants?

Additional "groundwork" tasks include organizing a design team and developing a work plan and schedule.

Establishing the base need not take a long time and can often be done largely at your desk or with the help of a few other people. We will revisit some of these tasks shortly.

Accomplishment 2: exercise development

Exercises, particularly tabletop, functional, and full-scale exercises, are developed by following an eight-step process:

1. Assess needs.
2. Define scope.
3. Write a statement of purpose.
4. Define objectives.
5. Compose a narrative.
6. Write major and detailed events.
7. List expected actions.
8. Gaining and announcing.

Accomplishment 2: Exercise Development (Continued)

These eight steps will be the focus of Unit 4.

> **Duplicate steps?**
>
> Performing a needs assessment, defining scope, and statement of purpose are important aspects of developing the overall exercise programme. They crop up again when establishing the base for an individual exercise. And here they appear as part of exercise development. Does this mean we just keep repeating the same steps?
>
> Certainly not. Sometimes (as in a simple exercise), the effort you put into these steps in the early stages will suffice and you will have already done the work when you get to development. Other times (as with a more complex exercise), you may need to come back to the needs assessment (or the scope or the purpose statement) and build on it, or refine it further for the particular exercise. The more complex the exercise, the more detailed the planning and development tends to be.

Accomplishment 3: exercise conduct

The day of the exercise is the culmination of your planning. What can you do to help ensure the exercise is successful? A few suggestions are listed below.

Conducting a successful exercise

- **Be clear.** The success of an exercise depends largely on the participants having a clear and consistent understanding of what is expected of them. Many exercises fail because the ground rules or simulation techniques to be used during the exercise are inadequately explained.
- **Sustain action.** Sustaining action through messages is one way to meet the key objectives of the scenario. Messages help keep all participants active throughout the entire exercise. This requires careful monitoring and control of the message flow throughout the exercise.
- **Foster realism.** Participants should be encouraged to treat the simulated emergency situations, e.g. the communications outages, damage, equipment failure, logistical limitations, personnel losses, as if they were actually occurring.
- **Establish timelines.** The establishment of valid timelines helps keep what is occurring in its proper order.
- **Review emergency call-off procedures.** To ensure that all participants understand the procedure for a real emergency call-off, they should be briefed before the exercise begins.
- **Capitalize on problem situations.** Situations that can cause stoppage in an exercise also have value because they place added stress on the system. Often they will test more effectively the organization's capability to cope in times of real emergencies.

Accomplishment 4: exercise evaluation and critique

An important part of any exercise is evaluating how well the objectives have been achieved. The objectives might relate to such issues as:

- needed improvements in the plan
- needed improvements in the emergency management system
- personnel training
- overcoming staffing deficiencies.

The extent and depth of the evaluation is determined by the participating organizations. Controllers' evaluations and observations may suffice for some exercises, whereas additional analysis by objective observers may be needed for others.

Critiques and reports analyze and explain the attainment or non-attainment of the exercise objectives and provide recommendations for addressing any deficiencies. Evaluators should be thoroughly familiar with the community's EOP and the area that they are evaluating.

Strategies for evaluating an exercise will be the focus of Unit 8.

Accomplishment 5: exercise follow-up

Follow-up is one of the most neglected areas of the exercise process. An exercise without evaluation and recommendations makes for an incomplete exercise. Similarly, recommendations without follow-up will keep you from getting the full benefit of the exercise.

How do you follow-up on an exercise? Selected strategies are provided below.

Exercise follow-up strategies

- **Assign responsibility.** Clearly assign tasks and schedules, and designate responsibility for each recommended improvement.
- **Monitor.** Establish a monitoring plan to track the progress of implementing recommended improvements.
- **Complete the cycle.** Build the testing of improvements into the next exercise. This is perhaps the surest way to make certain that they are implemented.

  It is probably not necessary to fully retest every objective. Rather, select a few recommendations that would illustrate improvements and include those in a future exercise.

This overview has presented the exercise process in a nutshell. As mentioned earlier, more detailed discussions of many aspects of this process will follow in later units. Now, let us take a closer look at some of the key tasks in establishing the base for an exercise. Other groundwork tasks—those that also become part of the exercise development process—will be the subjects of subsequent units.

Reviewing the current plan

The EOP describes how personnel in the organization should respond in the event of an emergency. It answers the question, "What resources, personnel, and procedures will be used to resolve problems created by an emergency?"

Your plan may have various names—perhaps one of the following:

- EOP
- Area Contingency Plan
- Operational or Operating Plan
- Contingency Plan

For simplicity, we will refer to all of these plans as **emergency plans**, and the plan is the place to begin.

Examining the plan will help you identify problems, select an exercise, define its purpose, and formulate objectives. While exercises test performance, it is the plan that describes the ideal performance. To use your plan properly as the base document, you must first become familiar with it.

What to look for in the emergency plan

While reviewing the plan, ask yourself key questions.

- What responses are currently planned (i.e. what hazards are the plans intended to address)?
- What resources, personnel, and procedures will be used to resolve problems?
- Are they different for various types of emergencies?
- Do roles vary according to the type of emergency?
- What training have response personnel experienced?
- What training is necessary?

Assessing capability to conduct an exercise

To design an exercise that simulates a real emergency, you must know what responses are planned and assess what capabilities are needed to meet those responses. Then you must ask, "Are we at a point where we can conduct an exercise?"

Before you launch into an exercise, it is important to find out if you have the resources—skills, funding, personnel, time, facilities and support. Deficiencies in any of these areas will have to be considered in the design of the exercise. When sheer enthusiasm says, "Let's do a full-scale exercise," sometimes the capability assessment will say, "We can only do a tabletop right now."

Questions to ask about capability and resources

The questions below, though very general, can help you assess your organization's level of capability. For example, you may find that before you consider planning an exercise, you will need to develop support and train people.

1. When was your organization's last exercise?
2. What exercise experience is available in your community or organization? What is your own experience? What is your staff's experience?
3. How much preparation time can you reasonably expect to have allocated to developing an exercise?
4. How much time can people devote to developing an exercise?
5. What skills can those people provide?
6. What physical facilities do you use when you conduct an emergency operation? Will they be available for an exercise?
7. What communication facilities and systems do you use in a real emergency? Will they be available for an exercise?
8. What attitudes do you expect of the chief executive and emergency service directors or other organizational leaders toward the exercise?

Addressing costs and liabilities

Costs and liabilities should be addressed early on.

Liabilities

A problem inherent in many exercises, particularly drills and full-scale exercises, is the possibility of personal injury or damage to equipment. Before planning the exercise, check the organization's or jurisdiction's insurance coverage.

Costs

Costs (both apparent costs and hidden costs) are incurred at every stage of exercise development. Plan ahead to be sure that you have the resources Some cost considerations are listed below.

Cost considerations

- Plan for a wide variety of costs. The following are a few examples, but the actual cost items will depend on the exercise and your organization's policies.
    - Staff salaries
    - Equipment and materials
    - Contract services
    - Miscellaneous items (e.g. coffee, pencils)
    - Fuel to run equipment and transport volunteers
- Below are examples of key questions to ask to avoid committing more resources than are available.
    - Will reimbursement for overtime be required if the exercise takes place on a weekend or evening?
    - If the exercise supports a hospital certification exercise, who will cover the costs?
- Employees should have their emergency management responsibilities reflected in their job descriptions.
- Time for participation in training, planning and exercising should be set aside by each employee with an emergency management responsibility.
- Costs for routine participation in exercises should be recognized by agency or organization officials.

Activity: getting ready for exercise design

The questions below relate to your organization's readiness for the exercise process. Write brief answers to the following questions in the spaces provided. (**Note:** this form also appears as Job Aid 3 in Appendix A.)

Self-assessment: resources and costs

1. **Plans**

   How familiar are you with the emergency plans, policies and procedures of your organization or jurisdiction?

   ☐ Very familiar
   ☐ Only general familiarity
   ☐ Familiar with only a portion
   ☐ Need to thoroughly review plans, policies and procedures

2. **Time**

   a. How far in advance would your organization realistically have to schedule to plan and design each of the following exercise activities effectively?

   - Orientation ______________________________
   - Drill ______________________________
   - Tabletop exercise ______________________________
   - Functional exercise ______________________________
   - Full-scale exercise ______________________________

   b. How much preparation time can reasonably be allocated to developing an exercise?

   - Actual person days:
   - Elapsed time to exercise:

Activity: Getting Ready for Exercise Design (Continued)

Self-assessment: resources and costs (continued)

3. **Experience**

   a. When was your organization's last exercise?

   b. What is your previous experience with exercises? (Check all that apply)

Orientation:	☐ Presenter	☐ Participant		
Drill:	☐ Controller	☐ Participant		
Tabletop exercise:	☐ Facilitator	☐ Participant		
Functional exercise:	☐ Controller	☐ Simulator	☐ Player	☐ Evaluator
Full-scale exercise:	☐ Controller	☐ Responder	☐ Evaluator	☐ Victim

☐ Took part in post exercise debrief

☐ Helped write an evaluation report

   c. What other exercise-related experience is available in your organization?

4. **Facilities**

What physical facilities do you use when conducting an emergency operation?

Will they be required for this exercise?	Yes ☐	No ☐
Will they be available for this exercise?	Yes ☐	No ☐

Activity: Getting Ready for Exercise Design (Continued)

Self-assessment: resources and costs (continued)
5. **Communications** What communication facilities and systems do you use in a real emergency? Will they be required for this exercise? Yes ☐ No ☐ Will they be available for this exercise? Yes ☐ No ☐
6. **Barriers** Are there any resource barriers that need to be overcome to carry out this exercise? Yes ☐ No ☐ If so, what are the barriers and how can they be overcome?

Activity: Getting Ready for Exercise Design (Continued)

Self-assessment: resources and costs (continued)
7. **Costs** a. What types of costs might be incurred for these exercises in your organization? Do not list exact figures, just types of expenses, such as wages and salaries, transportation. For an orientation: For a drill: For a tabletop exercise: For a functional exercise: For a full-scale exercise: b. Are there ways that different organizations can reduce costs (e.g. by combining exercises, cost-sharing, resource-sharing) and still fulfil programme requirements? Explain.

Gaining support

It is important, from the beginning, to establish authority for conducting the exercise. This means gaining the support of the highest possible official in your jurisdiction or organization.

Even if the chief executive does not participate in the exercise, his or her authority and support are essential. The top official can get nearly instant and complete cooperation from those who will participate in the exercise. It can be very difficult without that authority and support to put on an exercise involving departments, organizations, or agencies over which you personally have no authority.

Gaining support of the executive is not always easy, but the following approaches will help.

- **Gain support for the entire exercise programme.** Build a comprehensive, progressive exercise programme. The executive will be more receptive to an exercise that is part of a proven, consistent and goal-oriented programme than to an isolated exercise.

- **Protect the organization.** Make a conscientious effort to protect the organization or jurisdiction from lawsuits.

- **"Sell" the process.** Your needs assessment, capability analysis, purpose statement, and objectives are important building blocks for the exercise. Beyond that, they provide a valuable tool for selling the idea professionally—first to your boss and later to the chief executive of your organization.

Presenting the concept

Having the results of your early preparation activities in hand will add to your credibility when presenting the exercise concept to those in authority. Your presentation should include brief explanations of:

- the need for the exercise
- organizational capability (experience, personnel, costs)
- the type of exercise
- the scope of the exercise
- the purpose of the exercise.

Gaining Support (Continued)

- **Announce the exercise.** Broad support for the exercise may be gained in some instances by sending out an announcement. The announcement, often in the form of an exercise directive, should come from the chief executive, but you should be prepared to write it. The directive serves the purpose of authorizing you to conduct the exercise and giving you the clout you need to gain support from others.

Writing the exercise directive

The exercise directive will closely resemble the purpose statement (which will be discussed in more detail in Unit 4). The directive should contain the following:

- Purpose
- List of participating agencies, organizations, or departments
- Personnel responsible for designing the exercise
- Exercise date (or approximate dates)
- Point of contact for additional information

Note that the nature of the emergency and the location of the exercise are **not** revealed.

Regarding dates, unannounced exercises are not recommended. However, whether you specify the exact date will depend on the degree of surprise intended. At a minimum, a range of dates should be given. In selecting the exercise date, check the community calendar to avoid conflicting with a major event (e.g. an athletic event).

An example of an exercise directive is given on the next page. Although this example illustrates an Emergency Management Office announcement, a similar approach would be appropriate for an organization in the non-profit or private sector.

Sample exercise directive

24 February 20XX

TO: All Agency Directors
FROM: CHARLENE W. WILLIAMS
Chief Administrative Officer
SUBJECT: Emergency Exercise

A simulated emergency exercise involving a terrorist incident has been scheduled for sometime during the week of May 12–18, 20xx.

The purpose of the proposed exercise is to improve the following emergency operations.

1. Rapid assessment.
2. Notification and alert.
3. Scene isolation and perimeter control.
4. Mass casualty triage.

It is important that your agency participate in this exercise. We encourage involvement at the highest level.

I believe we all realize the importance of emergency exercises as a means to community preparedness. I fully support this exercise and intend to join with you in participating.

The Emergency Management Office will be coordinating the exercise. They will be contacting you to make necessary arrangements for the development and conduct of the exercise. For purposes of realism and interest, details of the exercise situation will not be made known prior to the exercise.

For further information, call Tom Smith at EXT 1234.

Assembling a design team

Planning an exercise requires a multitude of tasks, from designing the exercise to arranging detailed administrative matters. These tasks require the effort of a dedicated team and team leader.

Exercise design team leader

The exercise design team leader is responsible for the exercise throughout the entire development process—and for managing all administrative and logistical matters. In an exercise involving multiple organizations or jurisdictions, the team leader needs assistants or liaison persons from each entity to help coordinate the many details.

Who should lead?

Because the leader is so important, it is essential that this person be experienced and capable. The exercise design team leader should be someone who:

- can devote a considerable amount of time throughout the exercise cycle;
- is familiar with the emergency plan and has a sound understanding of the response organizations who will be participating;
- Is *not* a key operational member of one of these organizations. All key members should participate in the exercise, and they might not be able to participate fully if they are involved in the design.

What about the emergency manager?

Generally speaking, the person with chief responsibility for managing emergency events should be a player in the exercise. Therefore, instead of heading the design team, the emergency manager should assign someone else to develop and conduct the exercise. Prior to the exercise, the emergency manager should be given the same information about the scenario as other department heads, but no more.

Where staffing does not permit the emergency manager to assign someone else, he or she will need to play a dual role.

- Assume responsibility for developing the exercise and get people from other agencies to help.

Assembling a Design Team (Continued)

- Participate in the exercise, but on a limited basis. Despite being familiar with the exercise design, the emergency manager can still take a number of actions without compromising the realism, for example, negotiate and coordinate, support the decisions of the chief executive, and approve resource allocation.

Exercise design team

Exercise design is a complex task that should not be done by a single individual. The exercise design team assists the team leader in developing exercise content and procedures.

Design team responsibilities

- Determine the exercise objectives.
- Tailor the scenario.
- Develop the sequence of events and associated messages.
- Assist in the development and distribution of pre-exercise materials.
- Help conduct pre-exercise training sessions.

Ultimately, team members will be good candidates to act as simulators or controllers in a functional exercise.

Design team candidates

Ideally, the team should include a representative from each of the participating jurisdictions in a multi-jurisdictional exercise. In a single-jurisdictional or single-organization exercise, it should include representatives from the key departments. However, if this results in an unwieldy team, select a small core team that can draw on other members as needed.

Assembling a Design Team (Continued)

Backgrounds

Selecting team members with varied backgrounds helps with coordination and stimulates creativity. Some additional technical and administrative support may be required for typing, printing, and other mechanics involved in materials preparation.

Organizing the design team

Design teams are so varied—in number of members, team makeup (multi-agency or single organization), available skills, political climate, scope of the exercise programme to be developed, and many other factors—that there is no single approach to organizing a design team.

Generally, the team leader should use the kinds of teamwork strategies that make any team successful. Team leadership is too large a topic to address in this course, but there are many resources (courses, written documents and Internet resources) you can access to learn more about teamwork. A few suggestions for organizing the work of a design team are listed on the next page.

Assembling a Design Team (Continued)

Strategies for organizing the team

- Establish clear goals.
- Agree on a plan of action, with specific responsibilities and due dates, to meet the established goals.
- Reach consensus on a realistic schedule for completing milestone activities.
- Meet regularly to monitor progress.
- Work together to share expertise and resources to get the job done.
- Use the team interaction to learn more about each other's organizations. The more you know about other stakeholders, the better you will be able to work together to manage emergencies when they arise.
- Keep lines of communication open about new ideas, potential pitfalls, successes, problems, and needs. Shared information and creative problem solving are huge assets in emergency management.
- Use job aids to make the schedule, responsibilities, and progress visible and to keep everyone "on the same page."
- Some examples are provided in the following pages, but remember, these are only examples, and any job aid you use should be adapted to the needs of your team.

Assembling a Design Team (Continued)

Checklist. A simple "to-do list" similar to the following can be used provide an overview of the process and ensure that all main tasks are completed. (**Note:** this checklist appears as Job Aid 4 in Appendix A.)

Sample exercise development checklist

Mission

- ☐ ☐ Needs assessment
- ☐ ☐ Scope
- ☐ ☐ Statement of purpose
- ☐ ☐ Objectives

Personnel

- ☐ ☐ Design team
- ☐ ☐ Controller or facilitator
- ☐ ☐ Players
- ☐ ☐ Simulators
- ☐ ☐ Evaluators
- ☐ ☐ Management
  - ☐ Safety
  - ☐ Observers

Information

- ☐ ☐ Directives
- ☐ ☐ Media
- ☐ ☐ Public announcements
- ☐ ☐ Invitations
- ☐ ☐ Community support
- ☐ ☐ Management support
- ☐ ☐ Timeline requirements

Training/Briefings

- ☐ ☐ Train simulators, evaluators, controllers
- ☐ ☐ Players' pre-exercise briefing

Scenario

- ☐ ☐ Narrative
- ☐ ☐ Major/detailed events
- ☐ ☐ Expected actions
- ☐ ☐ Messages

Logistics

- ☐ ☐ Safety
- ☐ ☐ Scheduling
- ☐ ☐ Rooms/location
- ☐ ☐ Equipment
- ☐ ☐ Communications
  - ☐ Phones
  - ☐ Radio
  - ☐ Computers
- ☐ ☐ Enhancements
  - ☐ Maps
  - ☐ Charts
  - ☐ Other:

Evaluation

- ☐ ☐ Methodology
- ☐ ☐ Locations
- ☐ ☐ Evaluation forms
- ☐ ☐ Post-exercise debrief

After Action Documentation/ Recommendations

- ☐ ☐ Evaluation meeting
- ☐ ☐ Evaluation report
- ☐ ☐ Follow up ideas for next exercise

Assembling a Design Team (Continued)

Activities Schedule. For relatively simple exercises, a basic schedule may be used to show major activities of the team leader and the team and completion deadlines for each. An example is shown below. Deadlines should be realistic, based on team resources. (**Note:** see Job Aid 5 in Appendix A for a blank version of this schedule form.)

Sample activities schedule		
Deadline for completion	**Leader activities**	**Team activities**
3 months prior	▪ Hold initial planning meeting	
2½ months prior	▪ Brief government officials ▪ Arrange for facilities ▪ Determine simulation structure ▪ Convene and brief design team	▪ Attend team briefing
2 months prior	▪ Review and finalize scenario	▪ Develop/review exercise procedures ▪ Arrange simulation ▪ Arrange participation ▪ Review exercise scenario
1½ months prior	▪ Obtain exercise materials ▪ Prepare ideas for scripted messages	▪ Prepare participant information packet ▪ Prepare operational data
1 month prior	▪ Review messages with team	▪ Review messages with leader ▪ Review evaluation forms ▪ Print forms ▪ Prepare scripted messages
3 weeks prior	▪ Prepare briefing for participants	
2 weeks prior		▪ Integrate messages into time schedule ▪ Develop training sessions
1 week prior	▪ Prepare exercise facility	
2–4 days prior	▪ Conduct training session ▪ Train supervisors	▪ Assist in training sessions

Activities schedule

Sample activities schedule (continued)		
Deadline for completion	**Leader activities**	**Team activities**
Day of exercise	▪ Conduct participant briefing ▪ Perform pre-exercise check ▪ Supervise the exercise	▪ Assist with pre-exercise check
1 week after	▪ Help prepare draft of final report	▪ Review final report and make suggestions
2 weeks after	▪ Revise and submit report	
3 weeks after	▪ Submit recommendations	

Assembling a Design Team (Continued)

A more detailed plan is needed for a major exercise, which must be planned with the thoroughness of any major organizational effort. A Gantt chart, as shown in the example below, is a useful scheduling tool for such detailed planning.

A Gantt chart displays time across the top and a sequence of tasks down the left-hand side. Time can be given in days, weeks, or months. The duration of time devoted to each activity is represented by bars extending across the timelines.

The example shows task groupings (with supporting subtasks scheduled) and staff allocations indicated by initials in the left column.

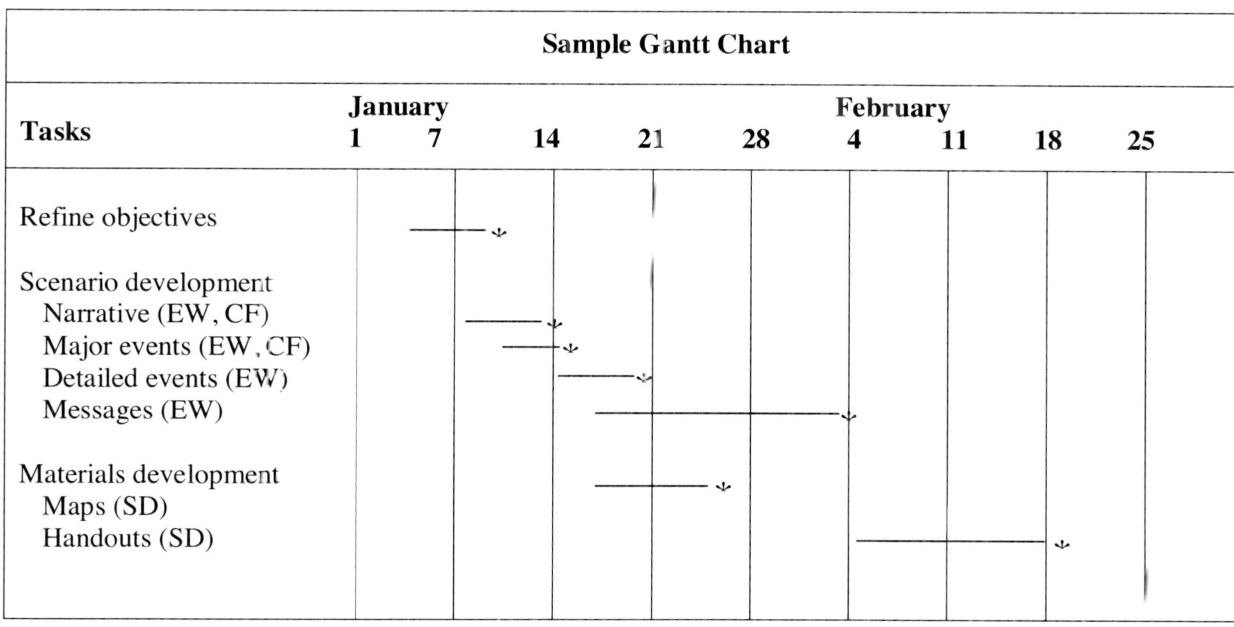

Activity: identify design team members

Who can assist in developing exercises? Consider your own staff, other departments or agencies, volunteer organizations, the United Nations Emergency Officer, or others as appropriate. Consider the following types of skills that may be needed for designing exercises:

- Planning
- Logistics
- Promotion
- Materials
- Scenarios/messages
- Other

Using the worksheet below, identify potential candidates for a design team to work on the exercise programme you outlined in Unit 2 and decide who could be the leader. For each person, identify what function or organization they represent and what they would bring to the design process (e.g. particular skills, experience, knowledge, perspective).

Design Team Worksheet		
Name	**Agency represented**	**Contributions/Qualifications**
Leader		
Members		

Exercise documents

Four major documents are developed during the exercise design process:

- Exercise Plan
- Control Plan
- Evaluation Plan
- Player Handbook

These documents are basically handbooks for specific audiences and serve as useful tools during exercise development, conduct, and evaluation.

Exercise plan

The exercise plan contains information that everyone needs and serves many purposes. For example, it can be used as per the following.

- To provide general information about the exercise for everyone concerned, including exercise overview, parameters, and timelines.
- As a guide for developers and participants.
- To assist participants in enhancing exercise play.
- As a promotional tool by the exercise director.

Control plan

The control plan is for controllers and simulators. It is not made available to players. It provides information about controller and simulator requirements and explains the exercise concept as it relates to controllers and simulators and provides the following.

- Establishes the basis for control and simulation of the exercise.
- Explains the procedures, responsibilities, assignments and support for exercise control and simulation.
- Defines the communications, logistics and administrative structure needed to support control and simulation during the exercise.

Exercise Documents (Continued)

Evaluation plan

The evaluation plan provides exercise evaluators, as well as controllers and simulators, with guidance concerning exercise evaluation procedures, responsibilities and support, and fulfils the following.

- Explains the exercise concept as it relates to the evaluation process.
- Establishes the basis for evaluation.
- Defines the communications, logistics, and administrative structure needed to support evaluation before, during, and after the exercise.

Player handbook

The Player Handbook provides exercise players with information needed to participate effectively in the exercise. This information is also discussed at the player briefings conducted prior to the start of the exercise.

Specific information included in each document is listed in the table on the next page. Templates of these documents are provided in the Exercise Tool Box.

Exercise Documents (Continued)

Contents of exercise design documents			
Exercise Plan	**Control Plan**	**Evaluation Plan**	**Player Handbook**
▪ Exercise type and purpose ▪ Scenario narrative ▪ Scope ▪ References ▪ Assumptions, artificialities and simulations ▪ Objectives ▪ Concept of operations ▪ Exercise management structure and responsibilities ▪ Safety and security ▪ Administration and logistics ▪ Exercise management job aids	▪ Exercise concept ▪ Pre-exercise player activity ▪ Assumptions, artificialities and simulations ▪ Concept for management, control and simulation ▪ Control team staffing ▪ Control team training ▪ Control team staff responsibilities ▪ Control team procedures ▪ Communications, logistics, administrative and other support	▪ Exercise concept ▪ Pre-exercise player activity ▪ Assumptions, artificialities and simulations. ▪ Concept for evaluation management ▪ Evaluation team staffing ▪ Evaluation team training ▪ Evaluation team staff responsibilities ▪ Evaluation team procedures ▪ Support for the evaluation team	▪ Exercise scope (concept of play, assumptions, artificialities and simulations) ▪ Scenario narrative ▪ Player procedures and responsibilities ▪ Safety and security ▪ Communications ▪ Reporting. ▪ Administrative systems ▪ Recommended pre-exercise training events ▪ Schedule of player exercise briefings. ▪ Provisions to review community plans, policies and procedures ▪ EOC (or other operating centre) procedures

Summary and transition

Unit 3 presented an overview of the main exercise activities and their relationships to one another. It also discussed some important aspects of laying the groundwork for an exercise, including:

- reviewing the current plan
- assessing capability to conduct an exercise
- addressing costs and liabilities
- gaining support
- assembling and organizing the design team.

Finally, this unit provided an overview of the four main design documents, the exercise plan, control plan, evaluation plan and player handbook. Unit 4 explains the eight steps in the design process.

For more information

http://training.fema.gov/EMIWeb/IS/is139lst.asp

Check **Knowledge check**

Carefully read each question and all of the possible answers before selecting the most appropriate response for each test item. Circle the letter corresponding to the answer that you have chosen.

1. The major task accomplishments in the exercise process are:

   a. Conducting tabletop, functional, and full-scale exercises.
   b. Developing, conducting, and evaluating the exercise.
   c. Establishing the base, exercise development, exercise conduct, critique and evaluation, and follow up.
   d. Conducting a needs assessment, writing objectives, and evaluating whether the objectives were achieved.

2. In conducting an exercise, unexpected problem situations are likely to lead to failure of the exercise.

   a. True
   b. False

3. Critiques and reports analyse how well objectives were achieved. They should also provide recommendations for addressing any deficiencies.

   a. True
   b. False

4. An important follow up strategy is to implement needed improvements revealed in the exercise and to test those improvements in the next exercise.

   a. True
   b. False

5. Availability of resources:

   a. Is irrelevant to the exercise process.
   b. Should not affect exercise design.
   c. Should be considered in scheduling and planning an exercise.
   d. Is seldom an issue for exercise design because exercises require few resources.

6. The exercise design team leader should be:

   a. The Emergency Manager.
   b. Someone who is familiar with the emergency plan.
   c. A key operational member of a participating organization.
   d. The chief official of the jurisdiction or organization.

Knowledge Check (Continued)

7. The design team should consist of:

   a. As many members as possible from all agencies or departments.
   b. Two or three members of the chief official's key staff.
   c. Members with varied backgrounds, representing key participating entities.
   d. The organization's training department.

8. The primary audience of the Exercise Plan is:

   a. Controllers and simulators.
   b. Participants.
   c. The facilitator.
   d. Everyone involved in the exercise.

9. The Control Plan explains the procedures, responsibilities, assignments, and support for exercise control and simulation.

   a. True
   b. False

10. The Control Plan should not be seen by ______________.

    a. Controllers
    b. Simulators
    c. Evaluators
    d. Players

11. The ______________ can be used as a promotional tool by the exercise director.

    a. Exercise Plan
    b. Control Plan
    c. Evaluation Plan
    d. Player Handbook

Answers

Knowledge Check (Continued)

1. c
2. b
3. a
4. a
5. c
6. b
7. c
8. d
9. a
10. d
11. a

Unit 4: Exercise Design Steps

UNIT 4

EXERCISE DESIGN STEPS

Introduction

Exercise design is much like scripting a play to ensure all players perform the correct actions and make the right decisions at the appropriate time. Tabletop, functional and full-scale exercises are based on a design process that includes the eight steps introduced in the previous unit.

1. Assess needs.
2. Define scope.
3. Write a statement of purpose.
4. Define objectives.
5. Compose a narrative.
6. Write major and detailed events.
7. List expected actions.
8. Prepare messages.

This unit describes what is involved in completing each of the eight steps.

Unit 4 objectives

After completing this unit, you should be able to list and explain the eight exercise design steps.

Step 1: assess needs

Some organizations plan exercises as a response to pressure or a "gut feeling." For example, someone may suddenly decide to do a full-scale exercise based on some dramatic disaster, because full-scale exercises generate a lot of excitement. Such hasty decisions usually result in failure and embarrassment. The best way to determine whether you need an exercise—and what kind of exercise is needed—is to study your situation systematically.

Conducting a needs assessment will help you define the problems, establish the reasons to do an exercise, and identify the functions to be exercised.

Begin with your plan

A needs assessment should begin with a review of the emergency plan and should address the following.

- Hazards—the risks that you are most likely to face and the priority levels of those hazards.
- Area(s) most vulnerable.
- Functions most in need of rehearsal.
- Potential participants (agencies, organizations, departments and personnel).
- Exercise requirements and capabilities.

If you have assessed your organization's exercise needs when planning a comprehensive exercise programme, you have a good basis already. For example, the assessment you completed in Unit 1 touched on many key issues. Consulting and updating that assessment will be an important step whenever a new exercise is considered for development.

Step 1: Assess Needs (Continued)

Lessons learnt

In doing a needs assessment for a single exercise, an obvious starting point is the evaluations of past exercises.

- Who participated in the exercise and who did not? Were only health agencies participating?
- To what extent were the exercise objectives achieved? Was everyone informed of each agencies roles and responsibilities?
- What lessons were learnt?
- What problems were revealed, and what is needed to resolve them? Was the chain of command and an official spokesperson identified?
- What improvements were made following past exercises, and have they been tested? Were plans subsequently revised?

Needs assessment results

In summary, your needs assessment should reveal the following types of issues if they exist.

- Primary and secondary hazards that the organization faces. The impact of a pandemic on agencies identified.
- Problems that need to be resolved.
- Problems that recur.
- Skills that need to be practised. Emergency notification and roles.
- Functions that are weak.
- Improvements implemented after previous exercises, which now need to be tested.
- New facilities, personnel, or equipment that have not been included in an exercise. Distribution of anti-virals?
- Weaknesses such as gaps, conflicting policies or vague procedures in the emergency plan or the SOPs.
- The need for role clarification. Have multi-agency organizational charts been created? Communication flow charts created and disseminated.
- The need for a certain type of exercise.

Step 1: Assess Needs (Continued)

A needs assessment form appears as Job Aid 1 in Appendix A. Now that you know a little more about exercising, you may wish to review your assessment from Unit 1 and update it if needed.

Step 2: define the scope

Defining the scope of an exercise means putting realistic limits on the issues that you identified in the needs assessment.

Why define the scope?

A needs assessment may reveal a wide array of concerns. Clearly you can't design an exercise that effectively practices:

- all functions
- in the context of all hazards
- using all agencies, organizations, or departments
- in all exercise formats
- employing all resources.

You will need to set priorities and make choices. It is important that the scope be clearly and narrowly defined.

Step 2: Define the Scope (Continued)

How is scope determined?

Many factors influence which areas of concern will be included in an exercise and which will not. Sometimes one decision will influence another (for example, the functions selected will determine who plays in the exercise). Other factors that help define the scope of an exercise include:

- expense
- availability of personnel and other resources
- seriousness of the problem
- capability of the exercise to address the problem
- skills and experience of the designers
- length of the exercise.

What does scope include?

There are five key elements of scope: type of emergency, location, functions, participants and exercise type.

- **Type of emergency.** An exercise is usually limited to one major event, although others, especially secondary events, might develop as the scenario develops. Hazards may be chosen for several reasons, including:
  - the emergencies that will generate the types of actions that needs to be practised, e.g. pandemic possible, pandemic imminent;
  - the highest priority hazards that the organization faces;
  - the hazards that haven't been exercised recently;
  - problems that have just recently developed.
- **Location.** Identify the location (a specific address) where the simulated event will occur. In pandemic scenarios, this could be an event that is occurring overseas, in a neighbouring country or within your country or locality. For tabletop and functional exercises, select a place where the hazard could realistically occur. In pandemics, it could begin anywhere. In a full scale rapid response and containment exercise, transportation problems or safety issues may make it necessary to make a compromise and find an area similar to the ideal location.

Step 2: Define the Scope (Continued)

- **Functions.** List the operations that the participants will practice. Be sure that the procedures within a certain function are clear and narrowly defined.

> **Example**
>
> To exercise a community's alert warning system, the following actions might be part of a response function.
>
> **Exercise Alert Warning System**
>
> - Notify the warning agency.
> - Electronic notification to staff or using media outlets to send out information. Notify Emergency Alert System (EAS) to interrupt programming with message.
> - Notify other sectors such as the police and military in your area.

- **Participants.** After the most important functions or needs have been identified, you can narrow the list of participating organizations and individuals to those that are required to carry out the actions.

> **Ask yourself –**
>
> - Which organizations need to be involved to carry out the function(s) being tested?
> - Which representatives from the identified organizations should be there?
>
> For example, in an EOC or other operations centre, you would typically want policy-makers, coordinators, and operations personnel. In an Incident Command Post (at the outbreak area), you would most likely want personnel who are knowledgeable in field operations and have some on-scene decision-making authority such as field epidemiologist, logisticians and others involved in containment efforts.

Step 2: Define the Scope (Continued)

- **Exercise type.** Finally, a decision must be made on the type of exercise.

> **Ask yourself –**
>
> - What exercises are most needed?
> - What experience have personnel had with the various types of exercises?
> - What stress level do we want?
> - What types of exercises are mandated by regulatory requirements?

After these issues have been settled, it is time to formulate them into a statement of purpose.

Step 3: Write a statement of purpose

The **purpose statement** is a broad statement of the exercise goal. It focuses and controls the whole exercise and includes the following.

- Governs the selection of the objectives, which in turn governs subsequent steps.
- Clarifies for your executive sponsor and potential participants why the exercise is being conducted.
- Is useful in communicating plans to staff, the media and community leaders.

Developing the purpose statement

A purpose statement is easily constructed. One approach is simply to incorporate the scope decisions (type of emergency, location, functions, organizations and exercise type) into a single sentence. A date is usually added.

Two sample formats are provided on the following pages.

Step 3: Write a Statement of Purpose (Continued)

Using the purpose statement as an exercise directive

Statement of Purpose sample 1

The purpose of the proposed emergency management exercise is to improve the following emergency **operations**.

a. Notifying staff and other agencies of the event (pandemic phase change).
b. Reduction of services, travel restrictions, evacuation warning.
c. Relocation or deployment or redeployment of staff.
d. Shelter management.

By involving the following **agencies** in a functional exercise simulating a pandemic situation.

a. Emergency management
b. Police, military, airport officials
c. Public works
d. Health departments, hospitals
e. Red Cross
f. Public schools

In country X on April 2.

Step 3: Write a Statement of Purpose (Continued)

Statement of Purpose sample 2

The purpose of the proposed emergency management exercise is to coordinate the activities of city and county government, volunteer organizations, and private industry in their response to a major incident; to provide training to staff; to test and evaluate the Alert and Warning, Evacuation and Shelter/Mass Care Annexes; and to enhance interagency coordination and cooperation by involving the following department or agency heads.

1. Medical Health Officer	Ministry of Health
2. Mayor	Local government
3. Emergency Manager	National
4. Emergency Manager	Local
5. Police chiefs	National and local
6. Law enforcement	Justice County Sheriff
7. PIO's	National, local media
8. Biology expert	WHO, universities
9. The Red Cross	Disaster Director
10. Hospital	Emergency Room Director

These entities will be tested on July 15, in a simulated exercise involving a potential or actual pandemic influenza situation in country or Province or Municipality X.

The exercise directive, as discussed in Unit 3, is a memo from the chief official in your organization or jurisdiction, sent to agencies or departments which support your needs. The directive is essentially a restatement of the purpose statement.

When using the statement of purpose as the basis for the directive, the following information should be added.

- Contact person and telephone number.
- Hours the exercise will be conducted.
- Exercise location (may be omitted to retain an element of surprise).

Activity

Activity: define exercise scope and purpose

In Unit 2, you outlined a comprehensive exercise programme for your organization. Select a tabletop or functional exercise from that outline. Then define the scope and write a statement of purpose using the worksheet (also provided as Job Aids 7 and 8 in Appendix A).

1. Highest priority hazards (major and secondary).

2. Geographic areas/locations of greatest vulnerability to these hazards, e.g. areas that do not have infection control measures.

3. Agencies/departments/organizational units. List below the entities that have a significant role in emergency management/response. Then enter check marks in any columns that apply.

Agency/Organization	Limited experience with major emergencies	New plans, staff, or organizational structure not yet exercised	Problems revealed in prior exercises

Activity: Define Exercise Scope and Purpose (Continued)

4. Types/levels of personnel you want to have in the exercise:

- ☐ Policy-making (elected officials, chief operating officers, department heads)
- ☐ Coordination (managers, EOC representatives, department deputies)
- ☐ Operations (field personnel, headquarters staff level)
- ☐ Public representatives (media, PIOs, general public)
- ☐ Other: ______________________________

5. Types of operations/functions that you want participants to engage in (e.g. notification, evacuation):

6. Degree of stress, complexity, time pressure the exercise should have:

	High	Medium	Low
Stress	________	________	________
Complexity	________	________	________
Time pressure	________	________	________

Activity: Define Exercise Scope and Purpose (Continued)

Exercise:
Scope:
Type of emergency:
Location:
Functions:
Organizations and personnel:
Exercise type:

Activity: Define Exercise Scope and Purpose (Continued)

Statement of purpose

Step 4: define objectives

Early in the development of an exercise, you must decide what the exercise is intended to accomplish. These outcomes, or objectives, must be specified clearly.

An **objective** is a description of the performance you expect from participants to demonstrate competence. Objectives go hand in hand with the purpose statement but are more specific and performance based.

Why define objectives?

Objectives are essential during the four stages of the exercise process.

- **Design process.** Objectives are the pivot point in the design process.
  - The needs assessment, scope, and purpose statement lead to the formulation of objectives.
  - The success of later actions and decisions begins with carefully written objectives.
  - The narrative, the major and detailed events, expected actions, and messages are all based on the objectives. In one sense, the objectives can be thought of as general statements of expected actions.

- **Exercise conduct.** During the exercise itself, elements of the exercise should be conducted according to the objectives to make sure that it stays on track.

- **Evaluation.** Writing objectives is the beginning of the exercise evaluation process. During the exercise, observers use the objectives to evaluate performance. After the exercise, the evaluation report is based upon those objectives. The process of identifying evaluation criteria takes place at the time objectives are written.

- **Follow-up.** During the follow up period, participants retain, plan and practice to address objectives that were not fulfilled.

Step 4: Define Objectives (Continued)

How are objectives determined?

Many objectives become evident at the time of the needs assessment, when designers identify problem areas. These needs can usually be translated into a statement of objectives.

> **Example.** Suppose your last exercise showed weaknesses in alert and notification, specifically a failure on the part of the EOC to analyze and implement call-down procedures. One of the resulting objectives would be to verify that the EOC is now able to notify the proper agencies according to the plan.

Objectives are also arrived at by breaking down a purpose statement into its logical components.

How many objectives?

There can be as few as two or three objectives in a small exercise, or as many as 100 in a large national exercise including many federal, state and local jurisdictions. For an average exercise, 10 or fewer objectives are recommended.

In larger exercises, each participating organization should be responsible for developing its own specific objectives, which are then incorporated into one exercise package by the design team.

What makes a good objective?

The main thing to remember about objectives is that they must be clear, concise and focused on participant performance. They should contain:

- an action, stated in observable terms;
- the conditions under which the action will be performed;
- standards (or level) of performance.

In other words, an objective should state **who** should do **what** under **what conditions** according to **what standards**.

Step 4: Define Objectives (Continued)

Here's an example.

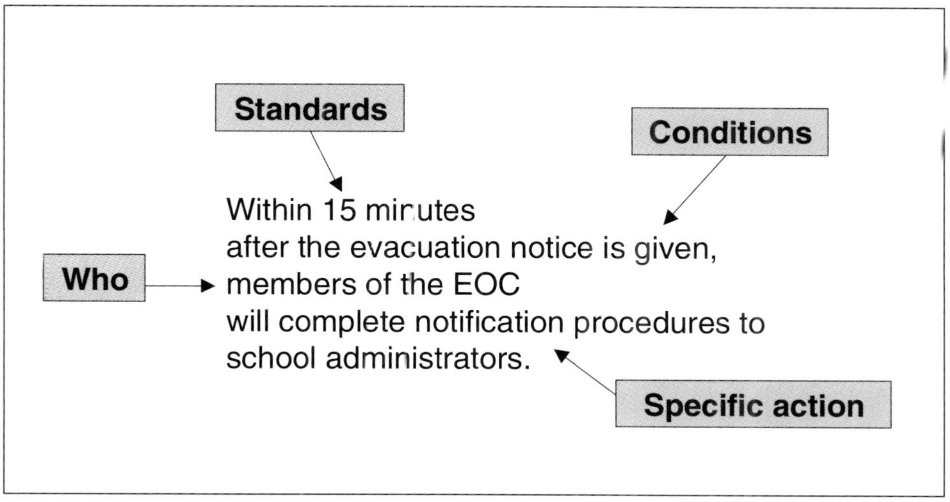

Writing SMART objectives

A useful guideline for writing objectives is the SMART system, which is described below. This system is easy to use and easy to remember.

SMART guidelines for useful objectives	
Simple	A good objective is simply and clearly phrased. It is brief and easy to understand.
Measurable	The objective should set the level of performance, so that results are **observable**, and you can tell when the objective has been achieved. This doesn't mean that you have to set a quantifiable standard. It just means that people can agree on whether they succeeded.
Achievable	The objective should not be too difficult to achieve. For example, achieving it should be within the resources that the organization is able to commit to an exercise.
Realistic	The objective should present a realistic expectation for the situation. Even though an objective might be achievable, it might not be realistic for the exercise.
Task oriented	The objective should focus on a behaviour or procedure. With respect to exercise design, each objective should focus on an individual emergency function.

Activity: analyze an objective

How good is the following objective? Use the provided rating scale to analyse the objective.

Objective: to demonstrate an understanding of the procedures necessary in protecting responder health and safety.

Is this objective?

Simple?	No	Needs Work	Yes
Measurable?	No	Needs Work	Yes
Achievable?	No	Needs Work	Yes
Realistic?	No	Needs Work	Yes
Task Oriented?	No	Needs Work	Yes

Is this a useful objective? Why?

Activity: Analyze an Objective (Continued)

Suggested response

The writer of this objective may have known exactly what he or she meant, but the statement does not work as an objective because it does not clearly communicate its meaning to anyone else. We are left with many unanswered questions, such as:

- What does "demonstrate" mean?
- What would the participant do to properly "demonstrate?"
- Who is the person who will demonstrate this understanding?
- What does "complete understanding" mean, and what standards would the participant have to reach to prove competence?
- What specific "procedures" does the writer have in mind?

Step 4: Define Objectives (Continued)

Points of review

Another way to ensure that the objective will provide useful measures is to include specific points of review—very specific items to be observed by an evaluator.

For example, the objective below is a little too vague to guide an evaluation team.

> **Objective.** Demonstrate the adequacy of displays and other materials to support emergency operations.

However, if the objective is supported by points of review similar to those shown below, it will be adequate:

Display	Yes	No
1. Status boards available in facility	☐	☐
2. Status boards used	☐	☐
3. Status boards kept updated by ________	☐	☐
4. Maps available	☐	☐
5. Maps up to date	☐	☐

Step 4: Define Objectives (Continued)

Helpful hints: word choice

- **Use concrete words.** One way to avoid vagueness is to use concrete words. Pay particular attention to the verb that describes participant performance.
- **Avoid vague verbs**, such as:
  - Know
  - Understand
  - Appreciate
  - Show the ability to
  - Be aware of
- **Use action words**, such as:
  - Assess
  - Clarify
  - Define
  - Determine
  - Demonstrate
  - Establish
  - Evaluate
  - Examine
  - Explain
  - Identify
  - Inspect
  - List
  - Notify
  - Operate
  - Prepare
  - Record
  - Report
  - Show
  - Test

Most objectives written for exercises describe performance, that is, the actions that a person or organization will carry out. Occasionally (especially in orientation seminars and some tabletops) the objective describes the understanding of a concept or a change in attitude.

Activity

Activity: compare good and poor objectives

Some examples of objectives are listed below. Review each example. Is it a good objective that will be useful to designers, players, and evaluators? Or is it a poor objective that needs improvement? Mark your answers in the provided boxes.

	Objective	Good	Poor
1.	Responders will establish an incident command post in the lobby of the high-rise structure within 15 minutes after the initial call for service.	☐	☐
2.	To get all of the emergency services to sit down in the same room together and talk.	☐	☐
3.	Demonstrate the ability to field a nuclear response team (fully equipped and assembled) within 30 minutes following a terrorist incident.	☐	☐
4.	Identify and activate an alternate communication system to be used as a backup within 30 minutes of failure of the primary communication system.	☐	☐
5.	To get the agencies to improve their disaster operations.	☐	☐
6.	The Mental Health Coordinator in the EOC will contact and deploy crisis intervention teams to the incident site after notification of request.	☐	☐
7.	To determine the capabilities of the fire/rescue department to effectively perform fire fighting, rescue, hazardous materials containment and similar hazard abatement duties during a major emergency.	☐	☐
8.	The warehouse manager will inventory and report to Central Processing the available stores of protective building supplies within 2 hours of notification by city officials.	☐	☐
9.	To identify the primary reason for slow response of ambulance units.	☐	☐
10.	Volunteers will be utilized.	☐	☐
11.	Emergency management staff will initiate and complete a call-back of EOC personnel as prescribed in the emergency plan.	☐	☐

Activity: Compare Good and Poor Objectives (Continued)

	Good	Poor
12. Demonstrate the ability of the Emmit International Airport Fire Brigade to respond to the farthest portion of the runway within 3 minutes of an alarm.	☐	☐
13. All facility personnel will respond properly to a chemical spill.	☐	☐

Activity: Compare Good and Poor Objectives (Continued)

Compare your answers with the following suggested responses.

	Good	Poor
1.	☑	☐
2.	☐	☑
3.	☑	☐
4.	☑	☐
5.	☐	☑
6.	☑	☐
7.	☐	☑
8.	☑	☐
9.	☐	☑
10.	☐	☑
11.	☑	☐
12.	☑	☐
13.	☐	☑

Activity

Activity: develop objectives

For the same tabletop or functional exercise for which you developed a statement of purpose earlier in this unit, write three objectives and identify the responsible organization. Use the SMART system, and remember to include:

- Action, stated in observable terms
- Conditions
- Standards

Exercise Objectives	
Objective #1:	**Organization:**
Objective #2:	**Organization:**
Objective #3:	**Organization:**

(**Note:** A similar worksheet appears as Job Aid 9 in Appendix A.)

Step 5: compose a narrative

An exercise is a scenario that simulates an emergency. Part of this scenario is the narrative, which is a brief description of the events that have occurred up to the minute the exercise begins. The narrative has two important functions.

First, it sets the mood for the exercise. Participants need to be motivated to participate. The narrative captures their attention and makes them want to go on.

Second, the narrative sets the stage for later action by providing information that the participants will need during the exercise.

Characteristics of a good narrative

- Is usually one to five paragraphs long.
- Is very specific.
- Is phrased in present tense.
- Is written in short sentences to lend immediacy and tension.
- May develop the situation chronologically (event with warning time).
- May emphasize the emergency environment.

For an emergency with warning time (e.g. a hurricane), the narrative often outlines the developing situation chronologically.

For an unexpected event (e.g. a chemical spill or terrorist bomb attack), the narrative may be shorter. Or, it may devote more detail to the environment of the emergency (e.g. a nearby school, other chemicals stored, rush hour approaching) to create intensity of feeling.

Step 5: Compose a Narrative (Continued)

Hints: outlining a narrative

You can outline a narrative by jotting down short responses (one or two words) to the following questions:

- What event?
- How fast, strong, deep, dangerous?
- How did you find out?
- What response has been made?
- What damage has been reported?
- What is the sequence of events?
- What time?
- Was there advance warning?
- Where does it take place?
- What are the relevant weather conditions?
- What other factors would influence emergency procedures?
- What is predicted for the future?

Then, when you're ready to write the text of the narrative, just take each of the key words and turn it into a brief sentence.

Two sample narratives are provided on the following pages. As you read the narratives, notice where the questions listed above are answered.

Step 5: Compose a Narrative (Continued)

Sample narrative: hurricane

The National Weather Service's National Hurricane Center issues news on the formation of a storm off the southern coast of the United States that appears to have hurricane potential. Tropical storm Anne is upgraded to Hurricane Anne and NWS issues a Hurricane Watch for a three-state area along the coast. Wind velocity and northwest movement over the last day have decreased, but an overnight change in direction to a steady northwest line calls for an immediate Hurricane Warning for five coastal counties of the state. Winds of 120 m.p.h. are predicted during the incoming tide, with high water expected to reach 12–15 feet over high tide. Low lying newly developed resort areas and heavy influx of visiting weekend campers have been advised to evacuate the area. Access bridges to barrier islands are narrow and could become impassable with 15-foot water heights.

Hurricane Anne, considered a very dangerous hurricane with high winds and an accompanying storm surge, will hit the coastal communities along Stevens Bay and farther inland, a population area of between 5000 and 25 000.

Following the hurricane watch, emergency service personnel notified elected officials and agency heads within the watch area. News media were also alerted and encouraged to broadcast the notice. When the warning of landfall within 24 hours was given, the Emergency Manager placed her staff on alert but did not activate the EOC. She has asked all appropriate emergency service personnel to meet at 07:30, approximately four hours after the warning was given. On its present course, the hurricane will make landfall at approximately 23:30. Flood stage from rising tides and tidal surge could, however, impact bridges by 16:00. All appropriate staff and emergency personnel are now gathered in the EOC.

Step 5: Compose a Narrative (Continued)

Sample narrative: air crash

A Boeing 747, en route from Panama to San Francisco, is experiencing in-flight engine problems and will have to make an emergency landing. Plans have been made to land at a large airport 200 miles north. However, the latest communication with the pilot indicates that the plane has lost engine power and is losing altitude too quickly to reach the large airport. Even though your city airport is too small to handle a 747, you are the only hope for the 350 passengers and 10 crew members.

Conditions at your airport are clear, and the surrounding area is dry. A hot, dry wind is blowing from the north.

The main runway lies along a relatively unpopulated suburban area. However, the likelihood of the pilots being able to control the huge plane and stay within the landing space is slim. The approach passes over populated suburban housing developments.

The airport control tower alerts its own crash/fire rescue units and requests that the local emergency services provide backup assistance in fire, police, medical, welfare and search and rescue capabilities.

Activity

Activity: outline a narrative

For the same tabletop or functional exercise as before, outline the key points of a narrative using the worksheet below (also provided as Job Aid 10 in Appendix A). Just list key words; you do not need to develop the full text at this time.

Narrative outline
Event:
How fast, strong, deep, dangerous:
How you found out:
Response made:
Damage reported:
Sequence of events:
Current time:
Advance warning:
Location:
Relevant weather conditions:
Other factors that would influence emergency procedures:
Predictions:

Step 6: write major and detailed events

Developing an exercise scenario is much like writing a play. In developing a play, the playwright organizes events into acts and scenes. Similarly, an exercise designer organizes events into major and detailed events.

Major and detailed events are occurrences, large or small, that take place after and as a result of the emergency described in the narrative. It may be helpful to think of them as **problems** requiring a realistic action that will meet exercise objectives.

The goal in developing events is to provide a structure that will:

- link the simulated event to the actions you want people to take;
- provide unity to the exercise; without the overall organization provided by major events, the exercise could dissolve into random actions.

Careful scripting is very important if you are going to produce a convincing, unified scenario rather than a series of unrelated, miscellaneous occurrences. It is also necessary for creating an exercise that is governed by objectives.

Developing major events

Major events are big problems resulting from the emergency. They should be likely events, based on case studies or operational plans, that call for realistic action.

Usually, the best way to arrive at a list of major events is to take it in two stages.

- First, identify several major occurrences—the high points in a sequence—that might follow the narrative events.
- Second, decide which of these events might generate situations that would test the objectives. Then concentrate on those that best support the objectives.

Step 6: Write Major and Detailed Events (Continued)

The major events in the following example were developed from the sample airplane crash narrative presented in Step 5.

Sample major events sequence for air crash scenario

1. Fuselage breaks apart and hits buildings below.
2. Jet fuel ignites several homes in the area.
3. About 60 survivors are thought to be trapped in the front section of the aircraft.
4. Several bystanders are injured on the ground.
5. A crowd convenes around the crash site.
6. Family members of victims begin to gather at the crash site.
7. Estimates of passenger casualties rise between 200 and 220 deaths and at least 70 severe burn victims.

Most events require action from one or more organizations. Because your goal is to develop an exercise that will test certain functions and organizations, the major events should be developed from your purpose statement.

Example

- Event # 1 above tests damage assessment and command and control.
- Event # 2 tests deployment of fire, police and medical services.
- Event # 3 tests search and rescue and medical personnel procedures.

Step 6: Write Major and Detailed Events (Continued)

Developing detailed events

Detailed events are specific problem situations to which personnel must respond. Each detailed event should be designed to prompt one or more expected actions for one or more organizations that are participating in the exercise. When the design task is small, it may not be necessary to distinguish between major and detailed events. However, for functional exercises, it is easier to write messages if you can base them on a list of detailed events.

There are several ways to develop detailed events. For example, you can:

- plan the detailed events and expected actions at the same time;
- work backwards, first identifying an action that you want players to perform, then listing a problem (a detailed event related to a major event) that would motivate the action;
- make a list of specific problems that are likely to occur in connection with each major event, then identify actions that would be expected as a result.

Whatever method you use, the result should be a list of specific events that are closely linked with actions that you want the participants to perform.

The detailed events in the following example were developed from event # 7 in the Sample Major Events Sequence ("estimates of passenger casualties rise between 200 and 220 deaths and at least 70 severe burn victims").

Sample detailed events

a. The mortuary is unable to accept the large numbers of remains resulting from the crash.
b. Local hospitals lack specialized facilities and personnel to treat large numbers of severe burn victims.
c. The American Red Cross has agreed to set up a family information centre to link victims and their families.

Step 7: list expected actions

Expected actions are the actions or decisions that you want participants to carry out in order to demonstrate competence. It is necessary to identify expected actions in order to perform the following.

- **Write messages.** Because the point of the exercise is to get the participants to think and react in certain ways, the script must be carefully developed to ensure that the messages get the planned results. Your list of expected actions will enable you to write effective messages.

- **Determine what should be evaluated.** The exercise evaluation will focus on whether the participants respond appropriately in an emergency. The list of actions will become the core of that evaluation.

Types of actions

There are four types of actions that the participants may carry out.

- **Verification:** gather or verify information.

- **Consideration:** consider information, discuss among players, negotiate, consult plan.

- **Deferral:** defer action to later, put action on priority list.

- **Decision:** deploy or deny resources.

How do you know what actions are appropriate in response to a given event? Refer to the emergency plan.

Step 7: List Expected Actions (Continued)

Relationship to objectives

Expected actions are closely tied to objectives. Objectives state general desired actions. **Expected actions** are a breakdown of objectives, the actions that would be taken by an organization or an individual to meet the objective.

The following example illustrates this relationship.

Example: objective and expected actions

Function Coordination and communication among the airport and the jurisdiction's emergency systems.

Objective Upon notification that a crash is imminent, response units will stage within 3 minutes, according to SOPs.

Event Landing of disabled aircraft is imminent.

Expected actions

Airport Control Tower

- Notify police, fire and medical personnel to proceed to airport.
- Alert hospitals of potential mass casualty incident.

Dispatch Centre

- Alert police, fire and medical supervisors.

Hospital

- Notify other medical facilities as appropriate.

Crash Fire Rescue

- Initiate Incident Command System.
- Notify dispatch of command post and staging locations.

Step 7: List Expected Actions (Continued)

Generally speaking, every detailed event results in one or more expected actions from various organizations. When you prepare a list of expected actions:

- list only those that involve the participating organizations (those identified in the exercise scope and statement of purpose);
- list expected actions for all exercise participants (however, it is not necessary that each detailed event generate responses from all participants).

Activity: write major and detailed events

First, review the objectives you developed earlier. With these objectives in mind, write two major events. Then complete the Expected Actions Planning Sheet that follows. (This sheet is also provided as Job Aid 11.)

Events
Major Event #1:
Detailed Events: **1.** **2.**
Major Event #2:
Detailed Events: **1.** **2.**

Note: Leave the objectives # column blank for now.

Activity: Write Major and Detailed Events (Continued)

For each detailed event listed on the previous page, identify the expected action, the organization responsible, and the objectives it addresses.

Expected actions planning sheet			
Detailed event	**Expected action**	**Organization**	**Objectives #**

(**Note:** See Job Aid 10 in Appendix A for a similar planning sheet.)

Step 8: prepare messages

Messages are used to communicate detailed events to exercise participants. One message may represent an event, or several messages may be needed to notify the participants of the event. Messages serve one purpose, to evoke a response, and cause exercise participants to make decisions and take actions that meet the exercise objectives.

In a full-scale exercise, the controller may input pre-scripted messages into the action. Participants receiving the messages make decisions or take action as they would in a real emergency.

Transmitting messages

Messages can be transmitted in various ways.

- Landline telephone
- Cellular telephone
- Radio
- In person
- Written note
- Fax

When transmitting messages in functional or full-scale exercises, try to use the method of transmission that would be most likely in an actual emergency.

Credibility

However messages are delivered, they must come from a credible source and be delivered through credible channels.

Example. If the participants are in the EOC, a message from John Q. Public to the EOC from an unlisted number would not be credible. However, the call could go to dispatch or to someone's secretary and the information could then be relayed to the EOC.

Step 8: Prepare Messages (Continued)

Relationship to expected actions

Messages have a direct relationship to expected actions. Each message is designed to generate one or more expected actions. The following example, begun earlier, lists some ideas for messages that could achieve the expected actions.

Example: objective and expected actions	
Function	Coordination and communication among the airport and the jurisdiction's emergency systems.
Objective	Upon notification that a crash is imminent, response units will stage within 3 minutes, according to SOPs.
Event	Landing of disabled aircraft is imminent.
Expected actions	<u>Airport Control Tower</u> ▪ Notify police, fire and medical personnel to proceed to airport. ▪ Alert hospitals of potential mass casualty incident. <u>Dispatch Centre</u> ▪ Alert police, fire and medical supervisors. <u>Hospital</u> ▪ Notify other medical facilities as appropriate. <u>Crash Fire Rescue</u> ▪ Initiate Incident Command System. ▪ Notify dispatch of command post and staging locations.
Possible messages	▪ Radio call from plane to tower. ▪ Tower calls police, fire and rescue. ▪ Plane requests runway be designated. ▪ Call from hospital requesting information. ▪ Calls to dispatch from media. ▪ Degrading radio communications with plane. ▪ Pilot feels major vibrations/noise on the plane.

(**Note:** Job Aid 12 in Appendix A provides a worksheet for planning messages to generate expected actions.)

Step 8: Prepare Messages (Continued)

Message variables

Messages, whether simple or complex, have four main variables, which are listed in the table below. Not all variables will be explicitly stated in every message, but they should be kept in mind as you write, because these variables form the classic definition of communication: who sends what to whom, with what effect.

	Message variables
▪ **Message source (WHO)**	Who sends the message (must be a credible source).
▪ **Transmission method (SENDS)**	How the message is transmitted (must be a credible means of transmission).
▪ **Message content (WHAT)**	Information conveyed. (Does the message contain the information needed by the recipient to make a decision?)
▪ **Recipient (TO WHOM)**	Who should receive the message. (Who would credibly receive it, and who ultimately needs to receive it in order to take action?)
All of these variables will influence the action taken (to what **EFFECT**).	

Some message examples are given on the next page.

Step 8: Prepare Messages (Continued)

Message examples

From: Police	To: EOC Police Staff
Cab door of trailer truck has been forced open and driver has been removed. Shipping papers indicate hydrochloric acid being transported. Acid flowing into sewers. Attempts to open rear trailer door ongoing.	

From: Environmental Protection (Field)	To: EOC Environmental Staff
Resident managers of apartments in area request information concerning safety of drinking water, water in swimming pools, and dwellings after evacuees are allowed to return to homes.	

From: EOC Fire Staff	To: Fire/Rescue Communications
Weather Service reports winds in an east-northeast direction at 10 to 15 mph with gusts of 20 to 25 mph. Forecast for continued rain with possible thunderstorms with strong gusting winds of up to 45 mph during the storm. Current temperature: 82 degrees.	

From: Chemical Facility Safety Officer	To: Fire Dept. HazMat Team Leader
The supervisor responsible for shutting off chemical valves in a plant area near the fire has not reported in. It is unknown whether the valves have been shut off and whether the person has evacuated.	

From: Betsy Ames	To: Township Fire PIO
My name is Betsy Ames. I'm a reporter for the Daily Express News. Can you give me a flood situation report for Hamilton, Jordan, Kemper, and the lakes in this area?	

Step 8: Prepare Messages (Continued)

Message format

Exercise designers often use a standardized form with spaces for the four variables, To, From, Method, and Content. A standard message form may also have spaces for message number, the time that the message is to be delivered, and the action to be taken. A sample messages format is shown below and appears as Job Aid 13 in Appendix A.

EMERGENCY EXERCISE

<MESSAGE>

TO:	METHOD	FROM:

NO:	TIME:

CONTENT: ____

____

____

____

ACTION TAKEN: ____

____

____

____

Step 8: Prepare Messages (Continued)

Composing a message

- In composing a message, begin with an expected action.
- Think about who could send a message and what that person could say to motivate the expected action.
- Keep it realistic.
- Think about the four message variables:
    - Who would credibly send the message?
    - How will the message be transmitted?
    - Who will receive the message? If not the decision-maker, where would the message be redirected?
    - Does the message provide all of the information needed to make a decision?
- Practice with the messages. Read them through with someone who is familiar with the organization involved. Does the message motivate the expected action? If it does, then you probably have a successful message.

Spontaneous messages

The majority of exercise messages in a functional exercise will be pre-scripted. And in the course of designing a functional exercise, it is a good idea to try to anticipate things that might go wrong and to provide the controller and simulators with ideas about ways to handle those situations.

However, in an actual exercise activity, the participants don't always respond as you expect them to. Then, it becomes necessary for the controller and simulators to improvise. Usually, the controller will decide on an appropriate response. But if the action is intense, the simulators may also have to make quick decisions and ad lib. If the controller and simulators are familiar with the scenario and objectives, their spontaneous messages can still fulfil the purposes of the exercise.

Activity

Activity: compose a message

Select one of the expected actions you listed in the last activity and compose a message that would generate the selected action. You can use the message form provided below. In your message, complete the following items:

- To
- Method
- From
- Content

EMERGENCY EXERCISE

<MESSAGE>

TO:	METHOD	FROM:

NO:	TIME:

CONTENT: ____

____

____

____

ACTION TAKEN: ____

____

____

____

Pulling it together: the Master Scenario of Events lIST

During a functional exercise, a Master Scenario of Events list is often used to monitor the progress of the exercise to keep it on schedule and on track. This chart (a list of events, time of occurrence, and expected actions) provides a picture of the whole exercise that is essential to the controller and helpful to simulators. The list should NOT be shown to the participants.

(**Note:** Job Aid 14 in Appendix A provides an MSEL worksheet. A partial example is shown below.)

Sample Master Scenario of Events list

Time	Message/Event	Expected Actions
07:35	Plane radios tower: losing engine power and altitude.	1. Tower notifies dispatch centre. 2. Dispatch alerts police, fire, medical to proceed to airport.
07:40 –07:50	Pilot reports major vibrations/noise; requests runway designation.	1. Tower designates runway; notifies dispatch of runway and potential for mass casualty incident. 2. Dispatch relays runway to police, fire, medical. 3. Dispatch notifies hospitals. 4. Crash fire rescue initiates ICS; notifies dispatch of CP and staging locations. 5. Dispatch relays CP and staging locations to police, fire, medical.
07:55	Hospital calls dispatch requesting more information.	1. Dispatch obtains potential number of casualties and relays to hospital. 2. Hospital notifies other medical facilities.
08:00	Media calls dispatch requesting information.	

Alternatives to self-developed exercises

In addition to developing exercises as described in this course, there are two other ways to prepare for an emergency management exercise.

- **Exercise-based training courses.** FEMA sponsors the Integrated Emergency Management Course (IEMC). Over a period of several days, this course accomplishes at least five important emergency management objectives: agency-specific briefings; mini-preparatory exercises; multi-agency, real-time exercise; creation of a teamwork environment; and participant critique.

- **Pre-packaged exercises.** Many state and local governments and FEMA have developed a wide variety of exercise packages that may reduce the design time and effort. The available exercise incidents include natural (e.g. weather-related) emergencies, hazardous materials incidents, terrorist scenarios, and others. These materials must be tailored to fit the specific needs and characteristics of your jurisdiction or organization. Therefore, you should review them carefully and adapt them before using them in your community.

Summary and transition

Unit 4 explained the eight-step design process. The next three units will provide more detailed information about designing and implementing specific types of exercises: tabletop, functional and full-scale exercises.

Reference Library

For more information

- Integrated Emergency Management Course:

  http://training.fema.gov/emiweb/iemc.htm

- Regional and state FEMA offices:

  http://www.fema.gov/about/regoff.htm

Check **Knowledge check**

Carefully read each question and all of the possible answers before selecting the most appropriate response for each test item. Circle the letter corresponding to the answer that you have chosen.

1. Which of the following is an example of a good exercise objective?

   a. To improve the number of accurate messages transmitted by the police to the communications centre.
   b. At the time the evacuation notice is received, the EOC policy and coordination groups will examine the needs of schools and other special facilities and organize notification according to SOPs.
   c. Proper procedures to declare a disaster or ask for outside aid will be taken.
   d. Interaction with other jurisdictions will be demonstrated.

2. An objective should specify the level of performance, so that results are observable.

   a. True
   b. False

3. Which of the following verbs would be most useful in an objective?

   a. Know
   b. Be aware of
   c. Appreciate
   d. Notify

4. The narrative

   a. Sets the stage for later action.
   b. Lists all of the events that will occur in the exercise.
   c. Provides the master scenario of events list and expected actions.
   d. Explains the objectives of the exercise.

5. The scope

   a. Sets the mood for the exercise.
   b. Establishes the limits of the exercise.
   c. States the objectives of the exercise.
   d. Lists the financial and human resources required to run the exercise.

Knowledge Check (Continued)

6. A convincing, unified scenario requires:

   a. 10 major events and approximately 100 detailed events.
   b. Messages written in a way that will not allow unexpected responses to occur.
   c. The players having knowledge of the events in advance.
   d. Careful scripting of events.

7. Every major event should require an action from all participants involved in the exercise.

   a. True
   b. False

8. Deferring action until later might be a valid action to take in response to a message.

   a. True
   b. False

9. In most exercises, the best way to deliver messages is in writing, using a standard form.

   a. True
   b. False

10. A ________________ is a useful tool to help the controller keep the exercise on track and on schedule.

    a. Message form
    b. Narrative
    c. Master Scenario of Events list
    d. Exercise directive

Answers

Knowledge Check (Continued)

1. b
2. a
3. d
4. a
5. b
6. d
7. b
8. a
9. b
10. c

Unit 5: The Tabletop Exercise

UNIT 5

THE TABLETOP EXERCISE

Introduction

Now that you have the "big picture" of the exercise process and the steps in designing an exercise, you're ready to take a closer look at specific kinds of exercises. This unit focuses on the tabletop exercise.

First, we'll review how a tabletop exercise works and the role of the facilitator. At some point you will probably be called upon to serve in this role, so we will discuss some guidelines for successfully facilitating a tabletop exercise. Finally, we will examine how the design steps learnt in the previous unit are applied to this type of exercise.

Unit 5 objectives

After completing this unit, you should be able to:

- describe the purposes and characteristics of a tabletop exercise
- describe the steps in facilitating a tabletop exercise.

Characteristics of the tabletop exercise

As learnt in Unit 2, a tabletop exercise simulates an emergency situation in an informal, stress-free environment. The participants, usually people on a decision-making level, gather around a table to discuss general problems and procedures in the context of an emergency scenario. The focus is on training and familiarization with roles, procedures, or responsibilities.

Characteristics of the Tabletop Exercise (Continued)

Purpose

The tabletop is largely a discussion guided by a facilitator (or sometimes two facilitators who share responsibilities). Its purpose is to solve problems as a group. There are no simulators and no attempts to arrange elaborate facilities or communications. One or two evaluators may be selected to observe proceedings and progress toward the objectives.

The success of a tabletop exercise is determined by feedback from participants and the impact this feedback has on the evaluation and revision of policies, plans and procedures.

Advantages and disadvantages

The tabletop exercise is a very useful training tool that has both advantages and disadvantages, as summarized in the following table.

Advantages and disadvantages of tabletop exercises	
Advantages	▪ Requires only a modest commitment in terms of time, cost, and resources. ▪ Is an effective method for reviewing plans, procedures, and policies. ▪ Is a good way to acquaint key personnel with emergency responsibilities, procedures, and one another.
Disadvantages	▪ Lacks realism and thus does not provide a true test of an emergency management system's capabilities. ▪ Provides only a superficial exercise of plans, procedures and staff capabilities. ▪ Does not provide a practical way to demonstrate system overload.

How a tabletop exercise works

In many respects, a tabletop exercise is like a problem-solving or brainstorming session. Unlike a functional exercise, problems are tackled one at a time and talked through without stress.

Problem statements and messages

A tabletop is not tightly structured, so problem statements can be handled in various ways.

- The facilitator can verbally present general problems, which are then discussed one at a time by the group.
- Problems can be verbally addressed to individuals first and then opened to the group.
- Written detailed events (problems) and related discussion questions can be given to individuals to answer from the perspective of their own organization and role, then discussed in the group.
- Another approach is to deliver pre-scripted messages to players. The facilitator presents them, one at a time, to individual participants. The group then discusses the issues raised by the message, using the EOP or other operating plan for guidance. The group determines what, if any, additional information is needed and requests that information. They may take some action if appropriate.
- Occasionally, players receiving messages handle them individually, making a decision for the organization they represent. Players then work together, seeking out information and coordinating decisions with each other.

Some facilitators like to combine approaches, beginning the exercise with general problems directed to key individuals and then passing out messages one at a time to the other players.

> **Handling problems.** It is usually wise to take the time to resolve problems, rather than hurry from one problem or message to the next, even though players sometimes will want to bypass the tough problems.

How a Tabletop Works (Continued)

Facilities and materials

It is recommended that the EOC or other operations centre be used for the tabletop exercise, for two reasons.

- It provides the most realistic setting.
- Needed plans, displays, and maps are available on the premises.

However, any conference facility that will comfortably accommodate the expected number of participants in a face-to-face setting will be adequate.

The number of participants and the scenario will determine the number and arrangement of tables for the exercise. Some facilitators like to arrange small groups around separate tables. Others prefer a U-shaped layout.

Provided reference materials should include emergency plans, maps and other reference materials that would normally be available in the EOC.

Facilitating a tabletop exercise

A tabletop exercise provides a relaxed environment of team problem-solving. Whereas functional and full-scale exercises are interactive, a tabletop is managed by a facilitator. The facilitator has a number of responsibilities.

- Introducing the narrative;
- Facilitating the problem-solving;
- Controlling the pace and flow of the exercise;
- Distributing messages;
- Stimulating discussion and drawing answers and solutions from the group (rather than supplying them).

Facilitating a Tabletop Exercise (Continued)

The facilitator must have good communication skills and be well informed on local plans and organizational responsibilities. Although the facilitator can be thought of as a discussion leader, the role can be much more. What follows are some guidelines for facilitating a tabletop exercise.

Setting the stage

The opening remarks and activities influence the whole experience. Players need to know what will happen and to feel comfortable about being there. Below are some guidelines for setting the stage for a successful tabletop exercise.

Guidelines for setting the stage

- **Welcome.** Begin by sincerely welcoming participants and putting them at ease.
- **Briefing.** Brief the participants about what will happen. This includes a clear explanation of:
  - ☒ purposes and objectives
  - ☒ ground rules
  - ☒ procedures
- **Narrative.** Start the exercise by reading (or having someone read) the narrative and introducing the first problem or message.
- **Ice breaker.** Try breaking the ice by beginning with a general question directed at one or two high-ranking officials or to the group as a whole. Later, other problem statements or messages can be addressed to other individuals or organizations.

Facilitating a Tabletop Exercise (Continued)

Involving everyone

It is important that everyone participates and that no one person or organization dominates the discussion. Tips for involving all of the participants are summarized below.

Ways to involve all of the participants

- Organize the messages so that all organizations must deal with a question or problem.
- Give extra encouragement to those who are a little reticent.
- Avoid the temptation to jump in with the right solutions when players are struggling. This will often hamper the discussion. Instead, try to draw out the answers from the players. They will be more likely to participate if they feel people are listening intently and sympathetically.
- Model and encourage the behaviours you want from the participants.
    - Give eye contact.
    - Acknowledge comments in a positive manner.

In-depth problem solving

The purpose of tabletop exercises is usually resolving problems or making plans as a group. That means going after real solutions—not superficialities.

Some facilitators make the mistake of trying to move too fast through the scenario, believing that they have to meet all of the objectives and get through all of the messages. However, that is not a good approach if nothing gets settled.

Remember, if you spend all the time on one big problem, maintain interest among players and reach consensus, then the tabletop is a success! Push players past superficial solutions. A few carefully chosen, open-ended questions can keep the discussion going to its logical conclusion.

Facilitating a Tabletop Exercise (Continued)

Controlling and sustaining action

To maintain a high level of interest and keep everyone involved, the facilitator needs to control and sustain the action. There are several ways to do this.

Ways to control and sustain action

- **Use multiple event stages.** Develop the scenario narrative in event stages. (For example, the initial narrative may involve warning, a later one could deal with search and rescue.) Then, as discussion begins to fade on one issue, introduce the next segment.

- **Vary the pace.** Add or delete problem statements and messages to alter the speed of the action. Occasionally give two messages at the same time to increase pace and interest.

- **Maintain a balance.** Maintain a balance between talking too long about a problem and moving along so fast that nothing gets settled. Don't hesitate to control the exercise tightly!

- **Watch for signs of frustration or conflict.** Always remember that the tabletop is basically training, not testing. People may come with fragile egos and little exercise experience. If you see mounting frustration or conflict, stop the exercise. Reach into your experience as a discussion leader to help the players resolve conflicts and feel comfortable.

- **Keep it low-key.** Avoid a bad experience by keeping in mind the low-key nature of the tabletop.

Activity

Activity: your ideas for facilitating a tabletop

You have read about many techniques for facilitating a tabletop exercise. But group facilitation styles and techniques are as varied as the facilitators who use them, and you may have some additional ideas about group techniques that would help you facilitate a successful tabletop exercise. Write down your ideas below.

Setting the stage:

Involving everyone:

In-depth problem solving:

Controlling and sustaining action:

Designing a tabletop exercise

The eight-step process outlined in Unit 4 is used to design a tabletop exercise.

1. Assess needs.
2. Define the scope.
3. Write a purpose statement.
4. Define objectives.
5. Compose a narrative.
6. Write major and detailed events.
7. List expected actions.
8. Prepare messages.

You can use the job aids provided in Unit 4. For a tabletop exercise, however, the process can be somewhat simplified. Because a tabletop is only partially simulated, it requires little scripting. The only roles are the facilitator, the participants (who respond in their real-life roles), and one or two recorders. Recorders take minutes and record decisions and usually do not need formal evaluation forms.

Applying the design steps

The first four steps are handled just as described in Unit 4. The remaining steps can be simplified as follows.

- **Narrative.** The tabletop narrative is sometimes shorter. It is nearly always given to the players in printed form, although it can be presented on TV or radio. When the purpose of the tabletop is to discuss general responses, the narrative can be presented in parts, with a discussion of problems after each part.

Designing a Tabletop Exercise (Continued)

- **Events.** The events should be closely related to the objectives of the exercise. Most tabletop exercises require only a few major or detailed events, which then can easily be turned into problem statements.
- **Expected actions.** A list of expected actions is useful for developing both problem statements and messages. It is always important to be clear about what you want people to do. However, in a tabletop, sometimes the "expected action" will be a discussion that will eventually result in consensus or ideas for change.
- **Messages.** A tabletop can succeed with just a few carefully written messages or problem statements. As always, messages should be closely tied to objectives and should be planned to give all participants the opportunity to take part.

  The messages might relate to a large problem (almost like an announcement of a major event) or a smaller problem, depending on the purpose of the exercise. Usually they are directed to a single person or organization, although others may be invited to join in the discussion.

Message example

General problem statement:

During a relocation process, what do you feel is needed for the support of your functions?

Specific message:

A call from Southside Nursing Home: they do not have enough transportation for all of their patients.

How many messages? It is a good idea to write a few more messages than you think you will need. However, if messages are carefully thought through, they will create a rather lengthy discussion. It's better to have 10 or 15 good messages than 20 or 30 hastily written ones.

(**Note:** a job aid for designing and facilitating a tabletop exercise is provided on the next page and also appears as Job Aid 15 in Appendix A.)

Designing a Tabletop Exercise (Continued)

Design

- ☐ Needs assessment, scope, statement of purpose, and objectives developed
- ☐ Narrative:
    - ☐ May be shorter
    - ☐ Presented all at once or incrementally
- ☐ Events:
    - ☐ Limited number
    - ☐ Presented as problem statements
- ☐ Expected actions:
    - ☐ May involve identification of appropriate responses, such as identification of gaps in procedures, reaching group consensus and developing ideas for change
- ☐ Messages:
    - ☐ Limited number (e.g. 10–15)
    - ☐ Involve everyone
    - ☐ Tied to objectives

Facilitation

- ☐ Welcome participants
- ☐ Briefing:
    - ☐ Purpose and objectives
    - ☐ Ground rules and procedures
- ☐ Narrative presentation (printed, verbal, TV, radio)
- ☐ Ice breaker questions directed at high-ranking officers
- ☐ Messages organized to involve all organizations
- ☐ Strategies to encourage the reticent
- ☐ Facilitate—don't dominate
- ☐ Model positive behaviours (eye contact, positive reinforcement)
- ☐ Aim for in-depth problem solving
- ☐ Strategies for sustaining action
    - ☐ Multiple event stages
    - ☐ Varied pace
    - ☐ Balanced pace
    - ☐ Conflict resolution
    - ☐ Low-key atmosphere

Designing a Tabletop Exercise (Continued)

Exercise examples

Examples of two approaches to tabletop exercises, "scenario development" and "single narrative with messages", are provided in the activities that follow. Review the plans carefully for similarities and differences.

A third approach is illustrated in the sample tabletop exercise provided in the **Exercise Design Tool Box** in the "Samples" Directory. In this exercise programme, local governments in a jurisdiction conduct concurrent tabletop exercises and report their actions via situation reports, Emergency Management Exercise Reporting System (EMERS) forms, and Participant Narrative Summary forms. The local exercises are followed by a regional exercise based on outcomes from the local exercises.

Activity: develop tabletop exercise problem statements

The following example illustrates one tabletop exercise design approach, the "scenario development" approach. The goal of this exercise (based on an earthquake scenario) is to generate general solutions to problems. To accomplish this, it presents the narrative in sections, with each section followed by a few problem statements. The facilitator would have participants discuss each problem statement until reasonable solutions have been reached, then proceed to the next narrative section and its accompanying problem statements.

Review the sample plan. Problem statements for the third narrative section have been omitted. In the spaces provided, develop two problem statements related to the third narrative section that would help achieve the objectives of the exercise.

Tabletop Plan: example 1 (scenario development approach)	
Organization:	Department of Management Services
Objectives:	1. Heighten awareness of city plan and standard operating procedures. 2. Identify and prioritize response activities. 3. Identify plans, policies, and procedures which are specific for this department and its various divisions. 4. Identify resources available to the department team. 5. Identify critical operations.
Participants:	<u>Department Level</u> ▪ Department of Management Services; Department Director ▪ Assistant to Mayor ▪ City Treasurer <u>Division Level</u> ▪ Division of Human Resources Management ▪ Division of Labour Relations ▪ Division of Accounting ▪ Division of Budget ▪ Treasurer ▪ Division of Purchasing ▪ City Recorder

Activity: Develop Tabletop Exercise Problem Statements (Continued)

Tabletop plan: example 1 (continued)

Scenario development 1 (narrative, part 1)

It is 9:00 a.m. on Tuesday, January 15. You have just felt a moderate earthquake as you worked at your desk in the City Office Building. You hid under your desk as plaster and glass flew around your office. After the shaking had stopped, you left the building from the nearest exit. Everyone is shocked, but apparently there are no serious injuries. You see several cracks in the rock fascia and wonder whether it is safe to re-enter the building. Your assistant comes up to you and tells you that he could not get a dial tone on his phone before exiting the building. Near panic, he asks these questions: "What do we do? How do we get help? How do we call the emergency command centre? How bad do you think the quake is?"

It occurs to you that these are questions that need to be answered.

Problem statements:

A. As division heads, what are our primary and secondary priorities and responsibilities? What are the time constraints on these activities?

B. As a Department Command Team, what are our priorities and responsibilities?

C. How is this department prepared to sustain division support/activity throughout a lengthy response? What about shifts for the command post members?

Scenario development 2 (narrative, part 2)

Because of possible structural damage to the building, the group decides to relocate the Department Command Post to ______________. It appears that the earthquake was not too severe; however, it will require a serious effort to keep the department services available and respond to the city's needs. Since the City Command Post is expecting the department to be located at the City Office Building in Roger Brown's office, it is necessary to discuss what to do and how to do it.

Problem statements:

A. If there is damage to this building, how do we select where we will relocate?

B. Roger Brown, at the City EOC, has to know of your whereabouts. How do you contact him at the City EOC to let him know where you are relocating? If radios are used, where did they come from? Are they operational?

C. During a relocation process, what do you feel as a team is needed for the support of your functions (e.g. administrative support, personnel and facilities)?

Activity: Develop Tabletop Exercise Problem Statements (Continued)

Scenario Development 3 (Narrative, Part 3)

It is now a day-and-a-half since the earthquake. None of you have been able to go home. Because of debris removal, all of the city response agencies, including public works, have been working almost non-stop. You hear of a street worker who was injured falling out of a tree while cutting down a broken branch. Also, last night a secretary complained that she was doing purchasing/contract procurement while only being paid as a secretary. She joked with her supervisor, but Frank was alerted to this concern by the supervisor for policy guidance.

Problem statements: (develop your problem statements in the spaces below).

A.

B.

Note: The scenario may continue through two or three more problem areas. The problem statements need not be addressed in sequence.

Activity: Develop Tabletop Exercise Problem Statements (Continued)

Suggested answers

A wide variety of problem statements could be developed based on the provided narrative section. Below are two examples. Yours will probably be different, but they should reinforce the exercise objectives given at the beginning of the example.

Examples of problem statements.

A. Does this body have the power to make policy decisions with regard to contract violation, shift problems for extended response periods, and injuries related to personnel performing functions outside their normal duties? Have liability issues been fully addressed?

B. What plan is in place to assist city responders (this group included) with family information assistance to reduce the stress of employees not knowing how their families are?

Activity

Activity: develop a tabletop exercise message

The following example illustrates another tabletop exercise design approach—a single narrative with messages. This plan focuses the participants on more specific responses to problems through the use of printed messages. The facilitator would present the entire narrative, then distribute the messages to specific players for discussion.

Review the plan. Imagine that **your organization** is included among the players. In the space provided, develop a message related to a **communications issue** that would affect your organization's ability to meet the stated objectives of the exercise.

Tabletop Plan: example 2 (message approach)

Objectives: Participants will demonstrate the ability to:

1. Coordinate effective evacuation.
2. Establish and maintain shelters for evacuees.
3. Coordinate round-the-clock field operations.
4. Ascertain safety levels needed to allow re-entry into affected areas.

Players:

- City/County Manager
- Emergency Manager
- Law Enforcement Representative
- Fire Services Representative
- Health/Environmental Representative
- Public Works Director
- Public Information Officer
- Volunteer Shelter Coordinator

Narrative: Steady rain has been falling in the region due to a moisture-laden low pressure system. Early into the storm, a Flood Watch was issued by the National Weather Service (NWS). Within 10 hours after the rain began to fall, the NWS issued a Flood Warning for low-lying areas along small streams. Some roads had to be closed and a small number of residents had to be evacuated along overflowing streams located northwest of the city.

Since the rain began, 16 hours have elapsed. Based on reports from designated citizens with rain gauges, total rainfall across the jurisdiction ranges from .1 inches to 4.7 inches. The soil has reached the saturation point, causing extensive runoff. Designated river spotters, water level gauges, and/or electronic river rise warning devices indicate that the river has risen to a dangerous level.

The 24-hour forecast calls for continued rainfall, with a clearing trend possible. Based on this data, the NWS issues a Flood Warning for areas along the river basin. Police report additional road closings due to high water, and a couple dozen more residents have been evacuated. Water levels are rising steadily, as a driving rain continues to fall. It is now 3:00 p.m. The EOC has been opened and the staff has been assembled.

Tabletop Plan: Example 2 (Continued)

Messages:

Message # 1 Time 4:15 pm
To: Public Works
From: Water Treatment Plant

Water continues to rise. River approaching flood stage. Debris is piling up at bridges. Water is now to top step of plant.

Message # 2 Time 5:00 pm
To: Police Chief
From: Police Officer

Motorists stranded in high water along Riverside Drive, south of bridge. They are on top of their cars in about 3 feet of water. The water is rising and moving faster.

Message # 3 Time 5:30 pm
To: Fire Chief
From: Battalion Chief

House on Nelson Drive contains 6 children under age 4, babysitter is only 11. They are in an upstairs bedroom. Water is over 2 feet deep in house. The sitter refuses to let the kids leave until the mother returns.

Message # 4 Time 6:45 pm
To: Public Works
From: Public Works Crew

Heart attack victim in River Oaks. Ambulance cannot get to him because of high water.

Message # 5 Time 7:30 pm
To: Dispatch
From: Citizen

Power line down near the Riverside City Park. One person appears dead, at least three others trapped in vehicles. Water is rising rapidly.

Message # 6 Time 7:30 pm
To: Emergency Manager
From: Dispatch

Switchboard and dispatch are overwhelmed. We are short of replacement personnel.

Message # 7 Time 8:45 pm
To: Volunteer Coordinator
From: Incident Coordinator

Emergency workers have been on job for many hours. They are in need of food and refreshments.

Message # 8 Time 11:00 pm
To: Public Information Officer
From: Public Utilities

Many parts of the city have lost power and telephone service, including many city agencies. Utilities do not know which agencies need repair first. Setting priorities is difficult.

Activity: Develop a Tabletop Exercise Message (Continued)

Tabletop Plan: Example 2 (Continued)

Message #__9__ Time 11:30 pm

To: City Manager
From: Incident Commander

Emergency personnel have been working for many hours and are exhausted.

Message #__10__ Time 1:00 am

To: Volunteer Shelter Coordinator
From: Red Cross

Shelters are running low on supplies, especially food. Evacuees have forgotten to bring vital medications.

Message #__11__ Time 3:30 am
To: Health/Environment Representatives
From: Policeman

Flood waters recede. Some residents are wanting to get back in their homes. Some lack proper ID. Some areas are still too dangerous.

Message #__12__ Time 9:00 am
To: Health/Environment Representatives
From: Citizen

There is a shortage of potable water.

Your Message:

Message # ____________ Time: _________________

To:

From:

Message:

Summary and transition

This unit provided an overview of the tabletop exercise, how it works, how to facilitate one, and how to design one using the eight design steps. Unit 6 will provide a similar overview of the functional exercise.

Reference Library

For more information

- FEMA's Emergency Management Exercise Reporting System (EMERS)

  www.fema.gov/onp/emers

FEMA: Descriptions of past tabletop exercises:

- Las Cruces counter-terrorism exercise: http://www.fema.gov/reg-vi/2001/r6_24.htm
- Pacific Northwest terrorism workshop: http://www.fema.gov/reg-x/2001/r10_36.htm
- Virtual tabletop exercise (Internet): http://www.fema.gov/pte/98365.htm
- Red River Basin flooding exercise: http://www.fema.gov/reg-viii/premac.htm
- Search www.fema.gov, "tabletop," for additional documents.

U.S. Department of Energy/Oak Ridge Associated Universities, Emergency Management Laboratory:

- Design course: *Beyond the Traditional Tabletop Exercise*: www.orau.gov/eml/manage.htm
- *What Would You Do If It Happened to You? A Tabletop Exercise Gives You a Chance to Find Out:* *www.orau.gov/eml/exercise.htm*

Check

Knowledge check

Carefully read each question and all of the possible answers before selecting the most appropriate response for each test item. Circle the letter corresponding to the answer that you have chosen.

1. A tabletop exercise:

   a. Involves a controller, players, and evaluators.
   b. Is an informal discussion guided by a facilitator.
   c. Requires a formal evaluation to determine its success.
   d. Must achieve all of the stated objectives to be considered a success.

2. A tabletop exercise:

   a. Provides a true test of an emergency management system's capabilities.
   b. Provides a practical way to demonstrate system overload.
   c. Is an effective method for reviewing plans, procedures and policies.
   d. Is better suited for field operations personnel than planners and policy-makers.

3. In a tabletop exercise, problems can be addressed to individuals, the group, or both.

   a. True
   b. False

4. It is more important to engage in in-depth problem solving than to meet all objectives and get through all of the problem statements.

   a. True
   b. False

5. If possible, a tabletop exercise should be run:

   a. At several field locations to simulate a real emergency.
   b. At a selected field location where an incident might occur.
   c. At a central location such as an operations centre.
   d. As a teleconference.

6. An important part of the facilitator's job is to:

   a. Maintain an even pace and consistent approach.
   b. Adhere to a highly structured agenda.
   c. Sustain action and keep everyone involved.
   d. Make sure the entire set of problem statements is discussed.

7. Most tabletop exercises:

   a. Require 50 to 100 detailed events.
   b. Are designed without identifying expected actions.
   c. Use general discussion in place of problem statements and messages.
   d. Require only a few major or detailed events.

8. The first four steps of the eight-step design process can be omitted when developing tabletop exercises.

   a. True
   b. False

9. Introducing multiple event stages is a good way to keep the action moving in a tabletop exercise.

   a. True
   b. False

10. A tabletop exercise should be:

   a. Low-key.
   b. Highly stressful.
   c. Formal and highly structured.
   d. Narrowly focused on field operations.

Answers

Knowledge Check (Continued)

1. b
2. c
3. a
4. a
5. c
6. c
7. d
8. b
9. a
10. a

Unit 6: The Functional Exercise

UNIT 6

THE FUNCTIONAL EXERCISE

Introduction

This unit focuses on the functional exercise. We will look closely at the characteristics of the functional exercise—how it differs from the tabletop, who participates, how it works, and key design considerations. This unit is pivotal, because later in the course you will develop a functional exercise based on what you have learnt here.

Unit 6 objectives

After completing this unit, you should be able to:

- describe the purpose and characteristics of a functional exercise;
- explain how designing a functional exercise differs from designing a tabletop exercise;
- describe the physical requirements and participant roles in a functional exercise.

What is a functional exercise?

The functional exercise simulates an emergency in the most realistic manner possible, short of moving real people and equipment to an actual site. As the name suggests, its goal is to test or evaluate the capability of one or more **functions** in the context of an emergency event.

It is important not to confuse "functional exercises" with emergency "functions." All exercises (tabletop, functional, and full-scale) test and evaluate functions contained in the Emergency Operations Plan (EOP). In this course, "functions" refers to actions or operations required in emergency response or recovery. The 13 functions recognized by FEMA and introduced earlier, in Unit 1, are:

- Alert Notification (Emergency Response)
- Warning (Public)
- Communications
- Coordination and Control
- Emergency Public Information
- Damage Assessment
- Health and Medical
- Individual/Family Assistance.
- Public Safety
- Public Works/Engineering
- Transportation
- Resource Management
- Continuity of Government

The key characteristics of functional exercises were discussed in Unit 2. You may wish to refer back to that discussion now. Below is a brief summary of the main points.

Key characteristics
▪ Interactive exercise, designed to challenge the entire emergency management system. Can test the same functions and responses as in a full-scale exercise without high costs or safety risks. ▪ Usually takes place in an EOC or other operating centre. ▪ Involves controller(s), players, simulators, and evaluators. ▪ Geared for policy, coordination, and operations personnel (the players). ▪ Players practice their response to an emergency by responding in a realistic way to carefully planned and sequenced messages given to them by simulators. ▪ Messages reflect a series of ongoing events and problems. ▪ All decisions and actions by players occur in real time and generate real responses and consequences from other players. Guiding principle: imitate reality. ▪ The atmosphere is stressful and tense due to real-time action and the realism of the problems. ▪ Exercise is lengthy and complex; requires careful scripting, careful planning and attention to detail.

What Is a Functional Exercise? (Continued)

Best uses

The functional exercise makes it possible to test the same functions and responses as would be tested in a full-scale exercise, without the high costs or safety risks. The functional exercise is well-suited to assess the following.

- Direction and control of emergency management.
- Adequacy of plans, policies, procedures, and roles of individual or multiple functions.
- Individual and system performance.
- Decision-making process.
- Communication and information sharing among organizations.
- Allocation of resources and personnel.
- Overall adequacy of resources to meet the emergency situation.

Activity: compare tabletop and functional exercises

In the following table, compare tabletop and functional exercises by writing a brief description in each of the cells.

	Tabletop	**Functional**
Degree of Realism		
Format/Structure		
Atmosphere		
Participants		
Who Leads		
Where Held		
Equipment Deployed		
Test Coordination		
Test Adequacy of Resources		
Test Decision-making Process		
Relative Complexity/Cost		
Formal Evaluation		

Activity: Compare Tabletop and Functional Exercises (Continued)

Suggested answers:

	Tabletop	Functional
Degree of Realism	Lacks realism	As realistic as possible without deploying resources
Format/Structure	Group discussion, based on narrative and problem statements/messages	Interactive; simulators deliver "problem" messages, players respond in real time
Atmosphere	Low-key, relaxed	Tense, stressful
Who Takes Part	Facilitator, participants (decision-making level); may use recorders	Controller, players (policy, coordination, and operations personnel), simulators, evaluators
Who Leads	Facilitator	Controller
Where Held	EOC, other operations centre, or conference room	EOC or other operations centre
Equipment Deployed	No	No
Test Coordination	Yes, on a discussion level	Yes
Test Adequacy of Resources	No	Yes
Test Decision-making Process	Yes	Yes
Relative Complexity/Cost	Small group; simple format; modest cost	Large scale; complex format; moderate cost to design and implement (higher than tabletop, lower than full-scale)
Formal Evaluation	No (self-assessment by participants)	Yes

Participant roles

As noted earlier, the functional exercise involves players, simulators, a controller, and evaluators. In a small jurisdiction or organization, one or two people may serve as controller, simulator, and evaluator. In larger jurisdictions, many more people will be necessary.

Let's take a closer look at what is involved in each role and how participants are selected.

Players

The players in a functional exercise are people who hold key decision-making or coordinating positions and would normally function in the operations centre.

By operations centre, we mean the central location that is designated in a real emergency for policy decisions, coordination, control, and overall planning. For a governmental jurisdiction, it would be the EOC; for a volunteer agency or private sector entity it would be the central location from which key decision-makers operate in an emergency situation.

Decision-makers. Key decision-makers would normally include leaders in government and key responding organizations: the mayor or other chief executive, and chiefs and coordinators of emergency services such as fire, police, EMS, Public Information Officer (PIO), and so on. In a nongovernmental organization, the CEO and other organizational leaders would participate.

Coordination and operations. Serving in the coordination and operations groups are people from various departments who work with policy-makers. In large exercises, a separate operations group carries out directives. In small exercises, the coordination and operations roles may be taken by the policy-makers.

The best guide in selecting who should participate in an exercise is the emergency plan.

Duties. The only job of the players is to respond as they would in a real emergency to the messages that they receive during the exercise. All of the decisions and actions of the players take place in real time and generate real responses and consequences from other players.

Participant Roles (Continued)

Simulators

In order to create a real-life environment, simulators portray the organizations that would normally interact with the players in the operations centre. They do this by delivering messages—descriptions of events or problems which require players to act.

> Some messages are scripted in advance; others are spontaneous responses to player decisions. They are input into the exercise by means of radio or telephone, or by written notes simulating radio and telephone transmissions.

Duties. Simulators are responsible for all actions taken by organizations or individuals outside of the EOC. They do the following.

- Send the players' pre-scripted messages representing private citizens, agencies, or other organizations, according to scheduled times in the sequence of events.
- Simulate all actions taken by an agency or other organization.
- Ad lib spontaneous messages as needed. Examples of times when a simulator may need to respond spontaneously include:
  - when a member of the operations centre issues a directive that results in events not anticipated in the scenario;
  - when a player asks for more information;
  - when a player decision is not logically linked to the next event in the scenario.
- Inform the controller of any deviations from the scenario, or special problems.

Once given directives, simulators are required to follow through and implement the directives in a professional manner.

Participant Roles (Continued)

Selection. Simulators must be able to ad lib intelligently in the situations just described, so it is important that they be familiar with the organization(s) they are simulating and with the sequence of events and messages. It is useful, therefore, to draw simulators from the organizations they will portray, and/or from the design team.

Numbers. It is difficult to give a rule of thumb concerning specific numbers of simulators needed for an exercise. The number of simulators will vary according to the following.

- Number of players
- Length of the exercise
- Knowledge and training of the simulators
- Communication channels available

For best results, try to have at least one simulator per organization represented in the operations centre, with extras to play the part of citizens or other private organizations.

Organizing. It is a good idea to group simulators according to function, in order to simplify the exercise and reduce the number of simulators needed. One approach is to organize them into three groups.

- Government agencies not participating in the exercise.
- Participating organizations: field units of organizations participating in the exercise (e.g. police, fire, public works) and private medical and support organizations.
- Other private facilities and individuals: citizens and nongovernmental organizations.

The following table illustrates how this approach could be used for a community.

Participant Roles (Continued)

Non-participating government entities	Participating organizations	Other private facilities/ individuals
One or two persons simulating: ▪ Federal regulators ▪ State or state area EOC ▪ County EOC ▪ Other city EOC ▪ State/Federal officers ▪ Care and shelter ▪ Resources and support	One person per organization simulating: ▪ City departments and agencies ▪ County departments ▪ Medical/health services ▪ Volunteer organizations	One or two persons simulating: ▪ Industries ▪ Commercial business ▪ Media ▪ Private citizens

Controller

The controller supervises the simulation or overall conduct of the exercise, making certain that it proceeds as planned and that objectives are reached.

The controller must be able to view the exercise as a whole and to think quickly on his or her feet. Players often make unanticipated decisions, and the controller must be able to respond to these.

Duties: The main duties of the controller are the following:

- Ensure that the simulators and evaluators are properly trained before the exercise.
- Orient the participants to the exercise and present the narrative.
- Monitor the sequence of events and supervise the input of messages, using the Master Scenario of Events List as a guide.
- Make decisions in the event of unanticipated actions or resource requirements.
- Adjust the pace of the exercise when needed—inserting more messages when it drags and discarding messages when the pace is too frantic.
- Maintain order and professionalism throughout the exercise.

Participant Roles (Continued)

Selection. Controllers can usually be drawn from the exercise design team. Because the team members are already familiar with the exercise, they are well suited to the task of keeping the exercise moving toward the anticipated conclusion.

Preparation. To properly prepare for the event, the controller should have the following items available.

- List of objectives
- Master Scenario of Events List
- Messages
- List of players
- List of resources available to the jurisdiction or organization

It is usually helpful to hold a briefing before the exercise to orient the staff members. At the briefing, the controller should train the simulators, ensuring that they are familiar with the scenario, objectives, resources, and the messages they will be responsible for delivering. The evaluation team leader should provide similar training to the evaluators, including exercise objectives, evaluator duties and schedule.

Evaluators

The evaluators observe the actions and decisions of the players in order to later report what went well and what did not. To do this, evaluators need to be familiar with the objectives, the exercise scenario, and the jurisdiction or organization that is undertaking the exercise.

Duties. Key duties of the evaluators include the following:

- Observing exercise progress and recording observations (usually on provided evaluation forms), taking care to remain unobtrusive in the process.
- Noting how well the exercise is fulfilling objectives and trying to identify problems if objectives are not met.
- Evaluating the actions of the players, not the players themselves. Documenting both positive and negative observations.

Participant Roles (Continued)

- Informing the controller during the exercise of any problems.
- Preparing brief written comments that can be included in the final evaluation and recommendation report that will be prepared by the emergency manager or other responsible party.

Unit 8 will provide more detailed information about the role of the evaluators.

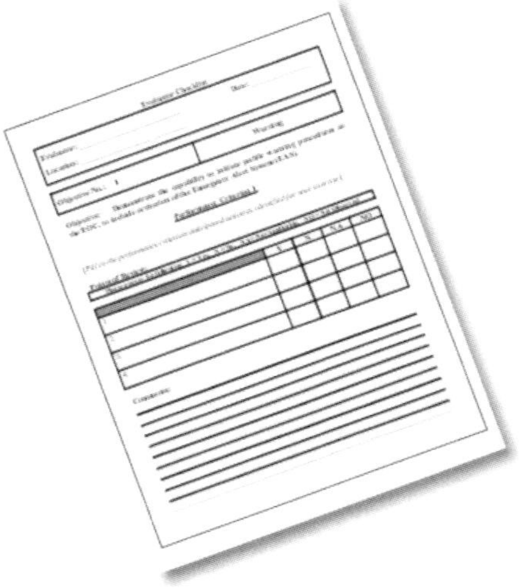

How a functional exercise works

A brief review of how a functional exercise works is given below. Keep in mind, however, that you will gain a better understanding of how a functional exercise works if you look for opportunities to observe one or, better yet, to participate in one.

The beginning

When a functional exercise begins will depend on its objectives. If testing the notification function is one of the objectives, then a "no-notice" exercise is useful. In this case, participants are given only the approximate timeframe scheduled for the exercise, anywhere from one day to several weeks. The exact time when it begins will be a surprise, allowing the exercise evaluators to observe how effectively notification and assembly at the command point take place.

In exercises where notification is not an objective, the exercise time is usually announced in advance.

How a Functional Exercise Works (Continued)

Briefing

Exercise participants may arrive on the scene of a functional exercise with only a vague notion of what is to take place. The exercise is much more likely to be successful if the participants receive a briefing that covers the following:

- Overview of objectives
- How the exercise will be carried out
- Time period to be simulated
- Ground rules and procedures

> **Keep the briefing short.** Avoid anything that distracts from the atmosphere of a real emergency. (For example, include a written announcement in the exercise materials to cover any administrative details such as restrooms and break times.)

Narrative

The exercise formally begins with the presentation of the narrative. It can be read aloud, presented on TV, computer, or slides; or dramatized.

Message delivery and response

The action begins as simulators and players interact with one another.

- Simulators communicate messages to players, and players respond as they would in a real emergency.
- Players make requests of simulators, and simulators react convincingly.

This ongoing exchange takes place according to the carefully sequenced scenario of events that governs what takes place, when each event occurs, and the messages used to inform the players.

How a Functional Exercise Works (Continued)

> **Example: message delivery/response**
>
> A message comes in from the incident site commander (a simulator) to the police chief (a player a real police chief). The message informs the chief of a traffic accident blocking emergency evacuation routes. The chief confers with aides, quickly plans a new traffic route, and telephones the incident site commander (simulator) with the instructions. The simulator carries out the instructions and reports back.
>
> Because the police chief might not react to the message as planned, simulators need to be prepared for a different response. They also must ensure that key events are kept active. For example, a player, not recognizing the importance of a key message, might delay action or fail to act. The simulator must then do something to cause the player to retrieve the event. If the situation reaches a point where the exercise cannot proceed until a decision has been made, the controller must force the issue.

Messages can arrive on paper, by telephone, by radio, or in person. Using telephones, where possible, increases the feeling of a real emergency, but whispered messages or written notes can also work well.

The success of the exercise depends on the extent to which the participants are able to carry out their functions as if they were in a real emergency. Exercise participants should be encouraged to think of each message as an actual event.

Encouraging spontaneity

The players should be able to decide among the full range of responses normally available to them during an emergency. Their ability to make decisions, communicate, or otherwise carry out their responsibilities should not be constrained by the exercise situation.

To allow the participants spontaneity, exercise controllers and simulators must be well trained and prepared to handle the unexpected. While this provides a better exercise for participants, it does place a burden on controllers and simulators who must be ready to "go with the flow" to some degree when the situation calls for it.

How a Functional Exercise Works (Continued)

Controlling the action

While simulators and players are transmitting messages and responding to them, the controller carefully monitors the interaction and progress.

Dealing with spontaneous decisions. The controller should be made aware of significant spontaneous decisions and make adjustments in the scenario where necessary.

> **Example**
>
> If a fire chief anticipated a later message by sending fire trucks into an area, the controller might need to stop a simulator from inputting a later message asking for fire trucks.

Adjusting the pace. The controller can control the pace of the exercise by adjusting the message flow—slowing things down when the pace is too frantic or speeding it up when the exercise drags. The controller can also even out the pace among participants. Remember, one inactive organization can distract others and bring down the intensity of the exercise. Avoid boredom by ensuring a smooth flow of messages.

Some specific suggestions for adjusting the pace are given on the next page.

How a Functional Exercise Works (Continued)

Ways to ensure a smooth flow of messages

- **Slow the pace by**:
  - **Rescheduling events** to allow more reaction time. Have the simulators wait before sending messages.
  - **Discarding messages** that are relatively unimportant or do not greatly impact other decisions. Throw away messages that don't contribute to the objectives.
- **Increase the pace and fill gaps by**:
  - **Speeding up the delivery pace** (varying from the planned schedule).
  - **Determining what is causing gaps** and being ready to add or alter messages spontaneously when needed. Look at organizations with gaps to see if they have been unintentionally ignored. If so, add messages. (It may be, however, that the organization simply has little to do during a particular period.)
  - **Keeping a supply of optional messages** on hand that can be added when needed.
  - **Adding side events**—routine actions a department would have to continue throughout an emergency. (For example, insert a routine traffic accident to put stress on police and fire departments. Report an unrelated heart attack to challenge medical personnel.)
  - **Adding secondary emergencies**—events that develop out of the main flow of exercise events. (For example, insert utility outages, water main breaks, gas leaks, media calls, and similar events to keep players involved between their own major events.)
  - **Adding special planning requirements** that would cause an inactive group to engage in a short-term preparedness activity (for example, have hospitals test emergency generators).
  - **Adding misdirected messages**—messages given to the wrong agency. Such messages can be used to gauge the agency's clarity of role definition and to test whether they forward the message properly.
- **Relieve overloads** on particular organizations by:
  - **Reassigning.** Verify that all messages are assigned to the right organizations, then reassign any messages that could be used by another organization.
  - **Thinning.** Divide the overloaded messages into two piles: (1) essential to the flow of the exercise and (2) nice to have. Then get rid of some from the latter group.
- **Maintain an even message flow** by maintaining a chart similar to the following:

How a Functional Exercise Works (Continued)

Planning an even flow of messages						
Check the times when messages are scheduled for delivery to each organization. Participating agency/organization						
	Fire	EMS	Public Works	EOC	Facility CEO	School
Exercise Start						
10:00	✔	✔				
10:03			✔			
10:06		✔		✔		
10:09	✔			✔		
10:12			✔	✔		✔
10:15		✔			✔	✔
etc.						

(**Note:** A blank planning chart is provided as Job Aid 16 in Appendix A.)

Skipping time

Functional exercises can depict events and situations that would actually occur over an extended time period (one or two weeks or more). In order to include multiple phases of the emergency (preparation, response, recovery, mitigation) in a two-day exercise, it would be necessary to stop the exercise periodically and advance the time by a number of hours or days.

These skip-time transitions should be kept to the minimum necessary to cover the scope of the exercise. They can usually be planned to coincide with a natural break point.

Who handles the time-skips? The controller is responsible for managing skip-time transitions and preparing transition updates to be presented to the participants before resuming the exercise.

Simulators are responsible for updating simulation displays to reflect the results of the previous events and participant actions. Actions that would have been undertaken during the transition period will be indicated as accomplished on the transition date.

The following table illustrates a skip-time schedule for a functional exercise.

How a Functional Exercise Works (Continued)

Sample skip-time schedule for a two-day functional exercise

Actual Time		**Period Simulated**		**Time/Activity Simulated**
Day 1	0900	Alert (mobilization)	3 hrs	First 3 hours (in real time) of Alert (mobilization)
	1200			
			SKIP	**Transition Statement**
	1300	Movement	3 hrs	First 3 hours (in real time) following evacuation order
	1600			
		ACTUAL TIME LAPSE OF 15 HOURS		
Day 2	0800			**Situation Update**
		Movement	3 hrs	3 hours (in real time) of movement
	1100			
			SKIP	Transition Statement
	1200	Sustaining	2 hrs	2 hours (in real time) of early sustaining period
			SKIP	**Transition Statement**
			1.5 hrs	1½ hours (in real time) of later sustaining period
	1530			

Facilities and materials

Location

Exercise where you operate. To the extent possible, the functional exercise should take place in the same facility and in the same operational configuration that would occur in a real emergency—usually the EOC or other operations centre.

A frequent objection to exercising at the operations centre is that there are not enough phones, or chairs, or restrooms. If that is the case, it is wise to find out in an exercise, not an emergency. If you cannot practice there, don't expect to be able to conduct an emergency response there.

Room arrangement

Various room arrangements can work for a functional exercise, depending on the size of the exercise. These are the basic requirements:

- Space for players—usually a table with plenty of work space;
- Area(s) set aside for simulators;
- Room for evaluators to observe;
- A place from which the controller can operate;

Small exercises. In very small exercises, a single room can work. The diagram below shows a simple layout for a small functional exercise.

Sample arrangement for a small functional exercise

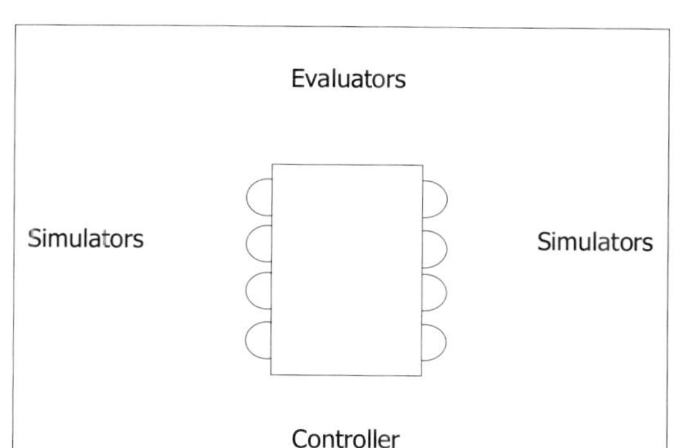

Facilities and Materials (Continued)

Complex exercises. The following layout would be appropriate for an elaborate functional exercise. Two rooms are shown, the simulation room and the operations centre, where the players are located.

Sample arrangement for a complex functional exercise

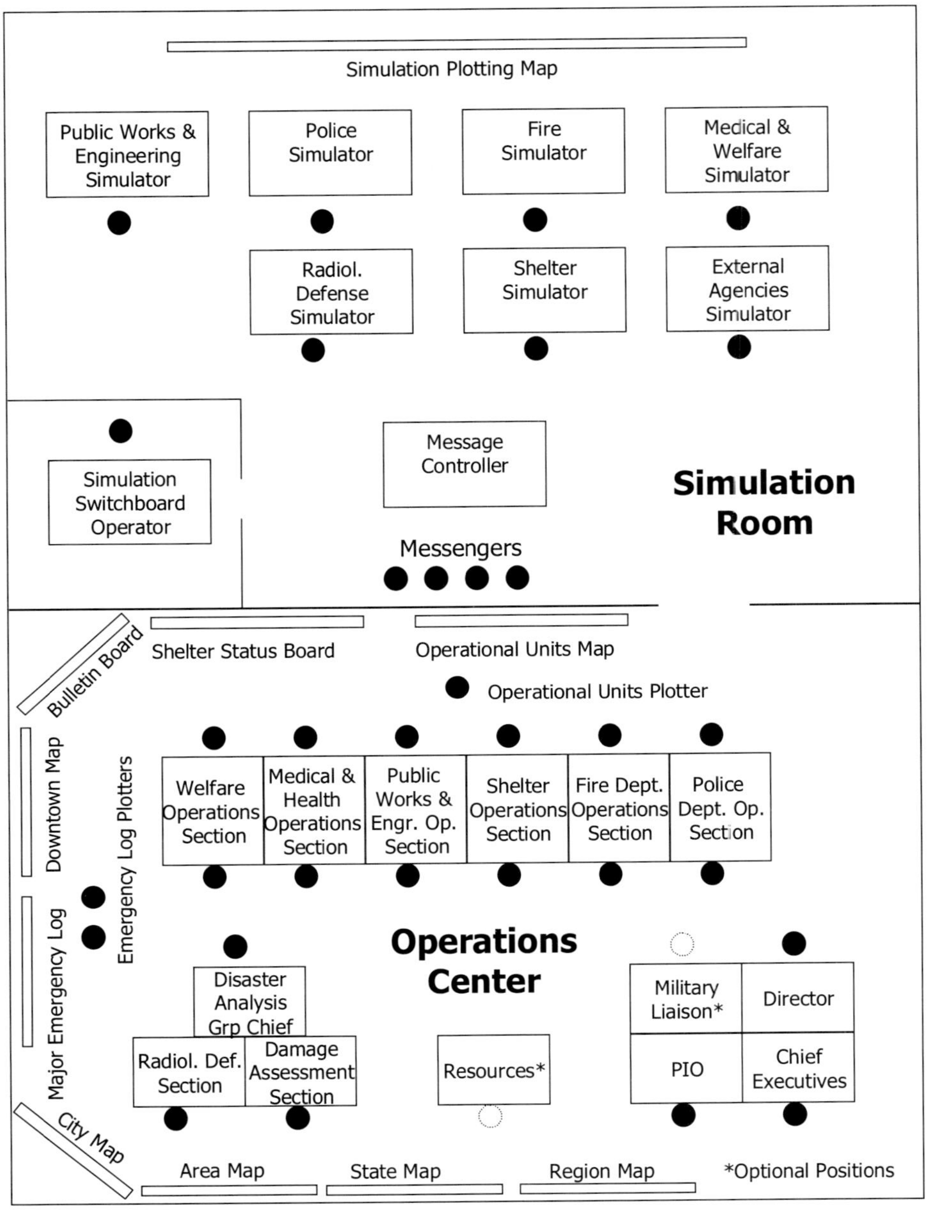

Facilities and Materials (Continued)

Simulation room. If more than one or two organizations or functions are being exercised, a simulation room is highly recommended. This room should comfortably house all of the simulators so that they can send, receive, and track messages and other communications with the players. It should be equipped with telephones or radios if they are to be used in the exercise. If message traffic is to be sent by hand, the situation room must be near the players.

Any layout should be adapted to the particular exercise and your local physical facilities. Whatever the layout, participant work spaces should be pre-designated and working supplies made available. (In the previous diagram, notice the work space assignments of the simulators and players.)

What about communications equipment?

Communications equipment is useful when full simulation is the goal. However, often it is wise to use both electronic equipment and written messages. When working in compressed time, it is easy for problems to arise concerning the following.

- Development of telephone banks for the simulators;
- Telephone overload for the players;
- Equipment breakdown;

For these reasons, some managers leave extensive use of communications equipment for a drill. In any case, the use of electronic communication should be carefully—and perhaps selectively—planned.

Equipment installation. When telephones will be the primary means of communication during an exercise, it may be possible to use existing phones. Or, it may be necessary to install special lines and extensions to provide the necessary communication links. In some facilities, where a central switching system is used, an operator may handle all calls.

Facilities and Materials (Continued)

Suggestions for successful communication links

- Prepare a special exercise directory of telephone numbers.
- Include communications procedures in the directory.
- If you don't have telephones, use a variety of other formats, such as the following:
  - Written messages
  - Simulated calls (sender whispers message in receiver's ear)
  - Hand signals (player who wants to call a simulator raises a hand to bring the simulator over)
  - Simulated speaker phone or radio (simulator speaks loudly to the players)
- If you use written messages, provide standardized message forms.

Displays and materials

Displays and materials, e.g. maps, charts, message forms and lists, are important in a functional exercise. In the room arrangement diagram for a complex exercise, notice the variety of maps available to the players. These materials are used to provide details for the scenario and keep track of activities.

Generally speaking, it's best to use what you use every day. An exercise is no time to get new maps and message forms. Test the ones that you are currently using.

Unit 9 will provide more detailed guidance about exercise enhancements, including communications equipment, displays and materials, and other matters.

On the next page is a checklist of facilities and materials suggested for a functional exercise. Depending on the scope and complexity of the exercise, specific items may or may not apply.

Designing a functional exercise

Design process

The full eight-step design process outlined in Unit 4 is used to develop a functional exercise. Whereas a simplified version of that process can be used to develop a tabletop exercise, a functional exercise—even a small one—requires careful attention to every step.

Exercise materials

The success of a functional exercise rests on a carefully scripted scenario package that includes the following.

- A convincing narrative.
- Major and minor events which grow out of the narrative and are carefully chosen to support the objectives.
- Arrangement of the events in a realistic and convincing sequence from the beginning to the end of the exercise.
- Expected actions which are tied closely to the objectives.
- A great number of specific messages (perhaps 100 or more in a larger exercise) that are so well conceived that players will respond with the expected actions.

Expect the unexpected

No matter how good you are at writing convincing messages, sometimes players will respond in unexpected ways. Although you should try to limit the unexpected as much as possible, occasionally a spontaneous reaction is better than the response prescribed in the emergency plan.

- A master scenario of events list that includes all of the messages/events, delivery times, and expected actions.

When you have completed the scenario package, you will use the developed materials to create materials for the exercise participants, including the Exercise Plan, Control Plan, Evaluation Plan and Player Handbook.

Designing a Functional Exercise (Continued)

Job aids and samples

Job Aids 7, 8, 9, 10, 11, 12, 13, and 14 are provided in Appendix A. They are the job aids introduced in Unit 4. They are well suited to the design of functional exercises. In addition, a Functional Exercise Checklist is provided on the following pages. This checklist (which also appears as Job Aid 17 in Appendix A) summarizes the special considerations for designing a functional exercise.

In Unit 10, you will have an opportunity to develop a functional exercise using similar instruments.

Designing a Functional Exercise (Continued)

Functional exercise design checklist: special considerations
Facilities and equipment: ☐ Sufficient work space for simulators and players ☐ Simulation room (if needed) near player room ☐ Space for message centre, control centre, observers (as needed) ☐ Clear work surfaces ☐ Communication equipment (telephones, switchboard) ☐ Parking ☐ Adequate ventilation and lighting ☐ Restrooms **Displays and materials:** ☐ Displays easily visible or accessible ☐ Maps (regional, state, local, area, downtown, operational units) ☐ Major events log, bulletin board, status boards, simulation plotting board ☐ Easels, chart paper ☐ Message forms ☐ Pencils/Paper ☐ Name cards **Beginning:** ☐ "No-notice" or scheduled (according to objectives) **Briefing (short):** ☐ Objectives ☐ Process ☐ Time period portrayed ☐ Ground rules and procedures **Narrative:** ☐ Verbal, print, TV, computer, slides, or dramatization ☐ Time-skips if needed **Messages:** ☐ Large number (depends on scope) ☐ Pre-scripted ☐ Optional pre-scripted for adjusting flow **Message delivery:** ☐ Written ☐ Phone ☐ Other (verbal, speaker phone/radio, hand signals) ☐ Simulators prepared for spontaneous message development ☐ Standardized forms for written messages **Strategies for adjusting pace:** ☐ Rescheduling ☐ Adding/Deleting messages ☐ Misdirecting messages ☐ Reassigning messages

Activity

Activity: identify functional exercise roles

For each of the following activities, indicate who has primary responsibility by placing a check mark in the appropriate column.

	Controller	Simulator	Player	Evaluator
1. Present the briefing.				
2. Observe and record exercise progress.				
3. Control the pace.				
4. Decide how to implement emergency plan procedures.				
5. Track progress.				
6. Ad lib in response to unplanned player actions.				
7. Decide how to handle unexpected situations in the exercise.				
8. Respond to events.				
9. Ensure that simulators and evaluators are trained.				
10. Coordinate with other organizations on joint responses.				
11. Deliver messages.				
12. Present the narrative.				
13. Inform the controller of deviations from the scenario.				
14. Act the part of organizations participating in the exercise.				
15. Act the part of organizations not participating in the exercise.				
16. Ensure that activities run smoothly.				
17. Supervise message input.				
18. Portray private citizens and facilities.				
19. Make decisions about departing from the planned event sequence.				
20. Compare exercise conduct to objectives.				
21. Update the situation board during skip-time transitions.				

Activity: Identify Functional Exercise Roles (Continued)

Suggested answers

	Controller	Simulator	Player	Evaluator
1. Present the briefing.	✔			
2. Observe and record exercise progress.				✔
3. Control the pace.	✔			
4. Decide how to implement emergency plan procedures.			✔	
5. Track progress.	✔			
6. Ad lib in response to unplanned player actions.		✔		
7. Decide how to handle unexpected situations in the exercise.	✔			
8. Respond to events.			✔	
9. Ensure that simulators and evaluators are trained.	✔			
10. Coordinate with other organizations on joint responses.			✔	
11. Deliver messages.		✔		
12. Present the narrative.	✔			
13. Inform the controller of deviations from the scenario.		✔		
14. Act the part of organizations participating in the exercise.			✔	
15. Act the part of organizations not participating in the exercise.		✔		
16. Ensure that activities run smoothly.	✔			
17. Supervise message input.	✔			
18. Portray private citizens and facilities.		✔		
19. Make decisions about departing from the planned event sequence.	✔			
20. Compare exercise conduct to objectives.				✔
21. Update the situation board during skip-time transitions.		✔		

Summary and transition

Unit 6 was the second of three units providing in-depth information about specific types of exercises. This unit provided information about the functional exercise, including key characteristics, participants, format, strategies for conducting the exercise and key design considerations. Unit 7 will discuss the full-scale exercise.

Reference Library

For more information

Many of the information resources cited in Units 1–4 also contain information about functional exercises.

Check **Knowledge check**

Carefully read each question and all of the possible answers before selecting the most appropriate response for each test item. Circle the letter corresponding to the answer that you have chosen.

1. The functional exercise:

   a. Simulates an emergency response in an actual field setting.
   b. Simulates an emergency situation in a relaxed group discussion.
   c. Simulates an emergency as realistically as possible without deploying people and equipment to the site.
   d. Simulates an emergency involving all of the functions, organizations, and personnel that would respond to an actual emergency.

2. The goal of a functional exercise is to test or evaluate the capability of one or more functions in the context of an emergency event.

   a. True
   b. False

3. An exercise that tested only notification procedures in response to a terrorist bombing would be:

   a. A drill
   b. A tabletop exercise
   c. A functional exercise
   d. A full-scale exercise

4. A functional exercise can test the same functions and responses as in a full-scale exercise without high costs or safety risks.

   a. True
   b. False

5. A functional exercise is similar to a tabletop exercise except that the functional exercise requires less scripting, planning, and attention to detail.

   a. True
   b. False

6. In a functional exercise, events are presented via problem statements or messages and then discussed by the group.

   a. True
   b. False

Knowledge Check (Continued)

7. A functional exercise is a good way to assess:

   a. Communication and information sharing among organizations.
   b. Response time of field personnel.
   c. Adequacy of response resources (personnel and equipment).
   d. Hazard analysis for developing the EOP.

8. _________ is the best location for a functional exercise.

   a. A field site similar to where an actual incident might occur
   b. The Mayor's office
   c. A 911 dispatch centre
   d. The Emergency Operations Centre

9. Key decision-makers in the jurisdiction or organization being exercised would normally assume the role of:

   a. Players
   b. Simulators
   c. Controllers
   d. Evaluators

10. Which of the following is NOT true of a simulator?

   a. He or she may deliver written messages.
   b. He or she is often called upon to rate the performance of key players.
   c. He or she may deliver messages verbally.
   d. He or she sometimes needs to make up a response to a player.

11. It may be necessary to delete planned messages if the pace of the exercise starts to drag.

   a. True
   b. False

12. When designing a functional exercise, the eight-step design process can usually be significantly shortened or simplified.

   a. True
   b. False

Answers

Knowledge Check (Continued)

1. c
2. a
3. a
4. a
5. b
6. b
7. a
8. d
9. a
10. b
11. b
12. b

Unit 7: The Full-Scale Exercise

UNIT 7

THE FULL-SCALE EXERCISE

Introduction

This unit focuses on the full-scale exercise. We will look closely at the characteristics of the full-scale exercise—how it differs from the other types of exercises, who participates, the role of the EOC, and key design considerations. At the end of the unit, you will develop an action plan for later use in designing a full-scale exercise for your organization.

Unit 7 objectives

After completing this unit, you should be able to do the following.

- Describe the purpose and characteristics of a full-scale exercise;
- Explain how designing a full-scale exercise differs from designing a functional exercise;
- Identify planning considerations for site selection and scene management for a full-scale exercise.

What is a full-scale exercise?

A full-scale exercise is as close to the real thing as possible. It is a lengthy exercise which takes place on location, using as far as possible, the equipment and personnel that would be called upon in a real event.

In a sense, a full-scale exercise combines the interactivity of the functional exercise with a field element. It differs from a drill in that a drill focuses on a single operation and exercises only one organization.

Eventually, every emergency response organization must hold a full-scale exercise because it is necessary at some point to test capabilities in an environment as near to the real one as possible.

However, there is more to a full-scale exercise than just practice in the field. As we discussed in Unit 1, various regulatory agencies have requirements for full-scale exercises which must be satisfied. In order to receive FEMA credit, for example, a full-scale exercise must fulfill three requirements.

- It must exercise most functions.
- It must coordinate the efforts of several agencies.
- In order to achieve full coordination, the EOC must be activated.

Activity

Activity: know your regulatory requirements

Answer the following questions about your own organization. If you are not sure of the answer, this question may require some research. You are encouraged to find the answers now, before continuing with the unit. Understanding your organization's requirements will provide an important foundation for the concepts covered in this unit.

1. What agencies or groups impose exercise requirements or guidelines on your organization?

2. What do they require concerning full-scale exercises? (Consider scope, frequency, numbers of organizations involved, coordination, communication, documentation, evaluation, or other issues.)

Key characteristics

The key characteristics of full-scale exercises were discussed in Unit 2. Below is a brief summary of the main points.

Key characteristics

- Interactive exercise, designed to challenge the entire emergency management system in a highly realistic and stressful environment.
- Tests and evaluates most functions of the emergency management plan or operational plan.
- Takes place in an EOC or other operating centre and at field sites.
- Achieves realism through:
    - on-scene actions and decisions
    - simulated "victims"
    - search and rescue requirements
    - communication devices
    - equipment deployment
    - actual resource and personnel allocation
- Involves controller(s), players, simulators (different from simulators in a functional exercise) and evaluators.
- Players represent all levels of personnel, including response personnel.
- Messages may be visual (e.g. staged scenes, made-up victims, props) and scripted.
- All decisions and actions by players occur in real time and generate real responses and consequences from other players.
- Requires significant investment of time, effort, and resources (1 to 1½ years to develop a complete exercise package). Attention to detail is crucial.

What Is a Full-Scale Exercise? (Continued)

The purpose of full-scale exercises

There are numerous reasons for conducting a full-scale exercise.

- Greatly expands the scope and visibility of the exercise programme.
- If well-planned, can attract public attention and raise credibility. (However, to be successful, it must be the culmination of a comprehensive and progressive exercise programme that has been developed as the organizational capacity has grown.)
- Is useful to test total coordination, not only among policy and coordination officials, but among field forces. At the same time, it can test inter-organizational coordination.
- Enables a jurisdiction or emergency management system to evaluate its ability to perform many functions at once.
- Can pinpoint resource and personnel capabilities and reveal shortfalls.

What is needed to run a full-scale exercise?

Some people wrongly believe that, once started, a full-scale exercise can run on its own steam. In fact, a full-scale exercise requires a substantial commitment of time, money, personnel and expertise and should not be undertaken without the necessary preparation. These are the most important requirements.

- Substantial experience with preparatory exercises of various kinds—drills, tabletops and functional exercises.
- Total commitment of all emergency service organizations.
- Support from the chief elected and/or appointed officials.
- Adequate physical facilities, including space for the EOC and field command posts.
- Adequate communication facilities (e.g. radios and telephones).
- Plans in place to handle costs (both evident and hidden), such as labour and time commitment.
- Carefully thought out and planned site and logistics.

Activity: compare functional and full-scale exercises

In the following table, compare functional and full-scale exercises by writing a brief description in each of the cells.

	Functional	Full-scale
Degree of realism		
Format/Structure		
Atmosphere		
Who takes part		
Who leads		
Where held		
Equipment deployed		
Test coordination		
Test adequacy of resources		
Test Decision-making process		
Relative complexity/cost		
Formal evaluation		

Activity: Compare Functional and Full-Scale Exercises (Continued)

Suggested answers:

	Functional	Full-scale
Degree of realism	As realistic as possible without deploying resources	As realistic as possible; resources deployed
Format/Structure	Interactive; simulators deliver "problem messages", players respond in real time	Interactive; simulators play roles at the scene, players respond
Atmosphere	Tense, stressful	Highly tense, stressful
Who takes part	Controller players (policy, coordination, operations), simulators, evaluators	Controller(s), players (all levels), simulators, evaluators
Who leads	Controller	Controller(s)
Where held	EOC or other operations centre	EOC and field site(s)
Equipment deployed	No	Yes
Test coordination	Yes	Yes
Test adequacy of resources	Yes	Yes
Test decision-making process	Yes	Yes
Relative complexity/cost	Large scale; complex format; moderate cost	Very large scale; highly complex; high cost
Formal evaluation	Yes	Yes

Full-scale exercise roles

Full-scale exercises involve one or more controllers, the participants, simulators, evaluators and a safety officer.

Controllers

One or more controllers manage the exercise. In some exercises, where there are multiple sites or organizations, there may be more than one controller. In this case, all of the controllers cooperate under the direction of a chief controller.

The controller (or chief controller) is responsible for ensuring that the exercise starts on schedule. The controller also designates an exercise control point from which all communications should be monitored.

Participants

A full-scale exercise involves all levels of personnel, including:

- **Policy-makers**—those who are responsible for making broad policy decisions. They might include the chief executive and his or her staff, the Public Information Officer, the emergency manager, key department heads and other elected officials.

- **Coordination personnel**—people from various departments who coordinate decisions of policy-makers and make plans for resources.

- **Operations personnel**—those who carry out the directives. Sometimes coordination and operations are the same.

- **Field personnel**—fire, police, EMS, search and rescue, volunteer groups, representatives of private enterprises who participate in the response and many others.

Full-Scale Exercise Roles (Continued)

Simulators

Simulators in a full-scale exercise are different from those in a functional exercise. In a full-scale exercise, simulators are the volunteers who pretend to be victims of the emergency event. For realism, they may wear makeup and they "act" injured, unconscious, hysterical, or dead—whatever the scene calls for.

Evaluators

Evaluators observe the action and keep a log of all significant events. This is important because so many of the actions will not be pre-scripted, but rather spontaneous responses to other actions. Evaluators may videotape exercise action.

Safety officer

There are so many potential safety issues in a full-scale exercise that a safety officer should be designated. This person's primary responsibility is to analyse the entire exercise from a safety perspective. We will return to safety measures a little later.

How the full-scale exercise works

Beginning

The full-scale exercise begins in a fashion similar to the functional exercise; whether it is announced or "no notice" depends in part on the objectives. The exercise designer will decide how and when the exercise is to begin. The trigger may simply be a call from dispatch, a radio broadcast, or a telephone call from a private citizen. The beginning for each participant should be as realistic as possible (that is, personnel should receive notification through normal channels).

Personnel from the emergency services that are taking part in the field component must then proceed to the assigned location, where a "visual narrative" is displayed before them in the form of a mock emergency to which they will respond.

Key decision-makers—those who would normally operate out of the EOC or command centre during an emergency—proceed to the EOC to fulfil their roles. Command posts are set up as required by the event.

How the Full-Scale Exercise Works (Continued)

Action

Actions in a full-scale exercise occur in the EOC, at one or more field sites, and at the related command posts. Actions taking place at the event site and command posts serve as input to the simulation taking place at the EOC.

Although medical personnel, hospitals, EMS, fire services, and other localized emergency operations do not usually require centralized command from the EOC, they do require coordination with officials at the command posts.

Sustaining action

Action is sustained by various means, including:

- Pre-scripted messages input by the controller(s);
- Messages and actions from the field that require action at the EOC;
- Spontaneous responses to the various messages and actions.

Command post messages

A field command post can be used as part of the message input into the EOC. Either the command post can be written into the scenario and have a set of pre-scripted messages to be transmitted by radio, or the command post controller can monitor the sequence of events and transmit spontaneous messages.

Exercise locations

Field sites. The main event site will depend on the exercise scenario and objectives. For example, if the central event involves a plane crash, the exercise might take place at an airport. A simulated terrorist attack could be staged in a public facility such as a convention centre or shopping mall. A hurricane or flood exercise might be dispersed over many locations.

Most events will involve additional locations such as secondary event sites, hospitals, mortuaries, shelters, and other support locations, and command posts will be established near the emergency sites. In fact, one of the reasons a full-scale exercise is so complicated is that activity is taking place in various locations, and all of the sites must be coordinated.

Emergency Operations Centre. Activation of the EOC or other operations centre is central to a full-scale exercise (just as it is in a functional exercise). The purpose of the EOC is to provide a policy and coordination facility for the Chief Executive Officer (CEO) and staff in order to respond effectively to an emergency.

In essence, the EOC is the voice of government during an emergency. Emergencies place strains on government—the demand for services escalates, while ability to deliver diminishes. Gathering information, making decisions, and directing necessary actions require close coordination between key officials. This coordination is best obtained if officials and support staff are in a centralized location with direct lines of communications.

The central location makes it possible to accomplish a number of tasks that would be impossible (or very difficult) separately.

- Information can be gathered, verified, and recorded in one spot.
- Officials can deploy resources in a timely and intelligent manner.
- Direction and control can be efficiently managed.
- Officials can coordinate actions and decisions.
- It's easier to set meaningful priorities when the key actors collaborate.

The size and makeup of the EOC differs according to the size of the jurisdiction or response system. The EOC may take up an entire floor of a building or a small room.

Designing the full-scale exercise

The design of a full-scale exercise can be quite difficult, often requiring the expertise of several response organizations. When developing your first full-scale exercise, it is usually advisable to start small and build to more complex exercises. Many of the potential difficulties relate to logistical problems, but others rest with the design.

The entire eight-step design process is used to design full-scale exercises, although they are applied somewhat differently than when designing tabletops and functional exercises. The differences arise from the fact that tabletop and functional exercises rely on **words** to obtain realism whereas a full-scale exercise—like a real emergency—gains its reality from **things.** There is a shift from a written scenario to visual reality represented by a real site, real people (some of them simulating victims), and real equipment.

Let's take a closer look at how each of the design steps is applied to designing a full-scale exercise.

The first four steps

As with any exercise, the first four design steps are:

1. Assess needs
2. Define the scope
3. Write a statement of purpose
4. Define objectives

For a full-scale exercise, these steps require deeper analysis and greater attention to detail. So much rests on the outcomes of these steps that you must be sure that you have created a clear picture of what is to be achieved through the exercise. If any of these areas is left at all vague, the problems will be greatly magnified later on.

Designing the Full-Scale Exercise (Continued)

Step 5: the narrative

The narrative is handled differently in a full-scale exercise. Since a lengthy verbal description is not needed to set the scene, the narrative is shorter.

Step 6: major and detailed events

Major and detailed events still exist in a full-scale exercise, but many of them exist as actual occurrences rather than as verbal descriptions. For example, an earthquake scenario may have to rely on words to simulate some events. However, other earthquake events can be simulated with fallen logs, bricks strewn around a building, dummies resting under beams, people acting injured or frightened and other props.

Even when events are presented visually, they cannot be random and haphazard. Each event must be carefully planned and staged to support objectives and generate the expected actions.

Step 7: expected actions

As with any exercise, expected actions must be specifically identified, based on the exercise objectives. A detailed list of expected actions is an important foundation for the exercise evaluation.

Step 8: messages

There are two kinds of messages in a full-scale exercise: visual and pre-scripted. Much of the action grows out of the initial message and scene set-up. The scene that is set up contains a number of "visual" messages to which participants need to respond. It may also be necessary to have some pre-scripted messages to move the action along.

For every event, try to anticipate all possible reactions. But it is important to be flexible enough to adapt to player actions and decisions that you had not expected. Sometimes an unanticipated response will be an improvement over the expected action.

If your scenes (events) are well planned, the exercise will go in the general direction you planned, even if a few of the decisions are off course.

Designing the Full-Scale Exercise (Continued)

> **Watch the details!**
>
> Make sure that the scenes you create are good enough to get the expected action. For example, if you have victims who don't know anything about medicine, either coach them in advance about their symptoms or tag them with symptoms and vital signs and apply makeup to simulate injuries. At the end of the exercise, you don't want players to say, "Well, I didn't know what you were getting at."

Special considerations

A full-scale exercise represents a huge logistical challenge, and it's easy to overlook details. One way to promote clear and creative thinking is to "walk the site"—either physically or mentally. By doing so, you may be able to identify potential problems and do more realistic planning. In mentally evaluating the scene, you should consider these points.

- Site selection
- Scene management
- Personnel and resources
- Response capability
- Safety and legal liability
- Emergency call-off
- The media

The following are some criteria and guidelines related to each of these areas.

Special Considerations (Continued)

Site selection

Selecting the right site for the exercise should be one of the first decisions you make. Because the exercise requires the mobilization of personnel and resources, space and realism are key.

Site selection criteria
Adequacy of space ☐ Is it large enough to accommodate the number of victims, responders and observers? ☐ Is there space for responders' and observers' vehicles? **Realism** ☐ Is the site as realistic as possible without interfering with normal traffic or safety? Examples: In simulating an overturned tanker truck on a main stretch of freeway, you can't tie up rush hour traffic for hours. You'll need to find a similar location to stage the crash. For a plane crash, try a secondary airport instead of a national airport. ☐ Is the type of emergency one that has a real possibility of occurring? (A credible emergency can elicit greater cooperation and participation.)

Scene management

Management of the scene refers to a number of issues, including:

- logistics at the scene
- creation of a believable emergency scene
- number of victims
- management of props and materials
- Number of controllers

The checklist on the following page includes questions you should consider related to scene management.

Special Considerations (Continued)

Scene management checklist
Logistics Where will players be set up? If there is a mobile EOC, where will they park? **Believability** How will you simulate the emergency? (e.g. if you will simulate fire, what will you use for smoke? How will you simulate a spilled chemical, broken glass or flood damage?) Who will serve as victims? (For greater realism, select victims from different age groups, with different body types and physical characteristics.) How will you ensure that victims realistically portray their roles? **Number of victims** How many victims does the type of emergency call for? What is the capability and capacity of hospitals and other resources to handle victims? What does the history of past events indicate about types and numbers of injuries? **Props and materials** What kinds of props and materials will be needed to simulate injuries, damage and other emergency effects (e.g. victims' makeup, dummies and construction materials)? **Controllers** How many controllers will be needed to manage the exercise sites? (In a multiple-site exercise, every site will require a controller.)

Special Considerations (Continued)

Personnel and resources

The scenario will help determine how many people (participants and volunteers) will be involved, how many and what kinds of equipment will be needed, and the potential costs. Consider the following factors in planning for personnel and resources.

- How many participants? (Sometimes it is necessary to scale down the exercise to a half day or less to increase participation.)
- How many volunteers?
  - scene set-up
  - victims
  - members of the public
- How many will need to be paid overtime?
- What kinds of equipment will be used?
- How many pieces of each type of equipment?
- How much fuel for vehicles and equipment?
- What kinds of materials and supplies will be needed, and how will they be obtained?
- Expenses:
  - overtime wages
  - vehicle and equipment fuel
  - materials and supplies.

Equipment

Keep your scenario reasonable in terms of equipment. True, you need to determine whether you have the resources to meet normal emergency demands. But don't get carried away. Hold your people to the use of **actual equipment**. Don't let them simulate use of equipment that doesn't exist.

Special Considerations (Continued)

Response capability

In planning both personnel and resources, take into consideration how the exercise might deplete the actual response capability of the organizations involved. It is unwise to compromise the ability to respond to real emergencies during an exercise. Here are some suggestions.

- Be sure there are enough personnel and resources to continue their responsibilities if a real emergency occurs. (In some cases, a call-off procedure will solve part of the problem.)
- Consider using second-shift personnel or mutual aid from other jurisdictions or organizations.
- Consider using volunteers as a smaller response shift.

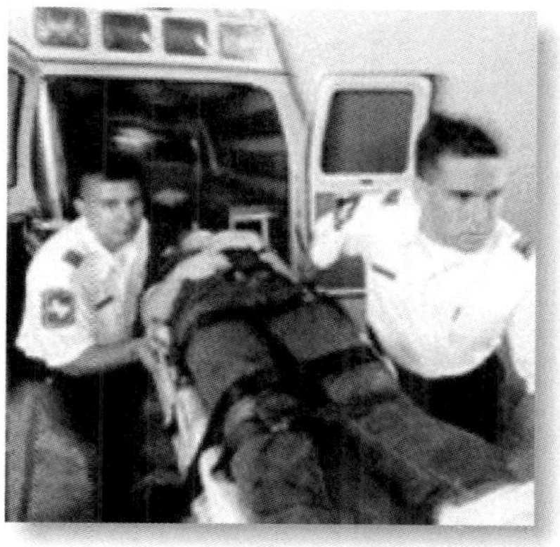

Safety

Total operation safety is an absolute must in a full-scale exercise. Awareness of safety issues must be considered throughout the planning and conduct of the exercise to ensure that safety problems are noted and eliminated. A safety officer should be designated whose primary responsibility is to analyse the entire exercise from a safety perspective. Suggested safety measures are on the next page.

Special Considerations (Continued)

Suggested safety measures
▪ Include safety as one of the activities in exercise development. ▪ Assign each exercise team member the responsibility of examining the exercise for safety within his or her discipline. ▪ Identify all possible safety hazards and resolve each one. ▪ Address safety as part of the pre-exercise briefing. ▪ Include safety factors in simulator and evaluator information packets. ▪ Examine each field location before the exercise to guarantee that safety precautions have been taken. ▪ Ensure that the safety officer has the authority to terminate an activity or even the entire exercise if a safety problem arises. ▪ Provide for call-off procedures in the event of an actual emergency.

Legal Liability

Legal questions of liability, including injuries during exercise conduct, must be researched by your local attorney.

Special Considerations (Continued)

Emergency call-off

A real emergency may occur during any exercise—especially a lengthy one. Be sure to keep enough personnel in reserve to handle routine problems.

In some instances, it may be necessary to stop the exercise to handle the real emergency. Every exercise should have a planned call-off procedure that will result in the prompt return of personnel and equipment to full duty status. The procedure should include a code word or phrase that the controller or safety officer can use to indicate that:

- the exercise has been terminated
- personnel should report to their regular duty positions
- all radio traffic will return to normal.

The procedures should be tested.

The media

A full-scale exercise of any magnitude will draw media attention whether you want it or not. If the exercise is well designed, favourable media reports are more likely. So include the media in your plans. They can be very helpful in gaining support for the program, and their presence will increase realism.

Also make allowances for observers and public information people. Decide in advance where you will locate them, and give them an opportunity to observe.

Job aids and samples

The job aids introduced in Unit 4 and provided in Appendix A, Job Aids 7 through 14, can be used to develop full-scale exercises. In addition, the checklist provided on the next page covers special considerations for full-scale exercises. (This checklist also appears as Job Aid 18 in Appendix A.)

Special Considerations (Continued)

Full-scale exercise planning checklist: special considerations

Participants:

- ☐ Controller(s)—sufficient to manage all event sites
- ☐ Simulators (mock victims)—different age groups, body types, physical characteristics
- ☐ Players (most functions, all levels—policy, coordination, operation, field)
- ☐ Evaluators
- ☐ Safety Officer

Site selection:

- ☐ Adequate space for number of victims, responders and observers
- ☐ Space for vehicles and equipment
- ☐ As realistic as possible without interfering with normal traffic or safety
- ☐ Credible scenario and location

Scene management:

- ☐ Logistics (who, what, where, how, when)
- ☐ Believable simulation of emergency
- ☐ Realistic victims
- ☐ Preparation of simulators to realistically portray roles
- ☐ Number of victims consistent with type of emergency, history of past events
- ☐ Types of injuries consistent with type of emergency, history of past events
- ☐ Victim load compatible with local capacity to handle
- ☐ Props and materials to simulate injuries, damage, other effects

Personnel and resources:

- ☐ Number of participants
- ☐ Number of volunteers for scene setup, victims, etc.
- ☐ Types and numbers of equipment
- ☐ Communications equipment
- ☐ Fuel for vehicles and equipment
- ☐ Materials and supplies
- ☐ Expenses identified (wages, overtime, fuel, materials and supplies)

Response capability

- ☐ Sufficient personnel kept in reserve to handle routine non-exercise events

Safety

- ☐ Safety addressed through development
- ☐ Each design team member responsible for safety in own discipline
- ☐ Hazards identified and resolved
- ☐ Safety addressed in pre-exercise briefing, simulator and evaluator packets
- ☐ Each field location examined for safety issues
- ☐ Safety officer designated, given authority

Legal liability

- ☐ Legal questions of liability researched by local attorney

Emergency call-off

- ☐ Call-off procedure in place, including code word/phrase
- ☐ Call-off procedure tested

Media

- ☐ Role of media addressed in planning, used as a resource to gain favourable exposure
- ☐ Media and observers considered in logistical planning

Activity

Activity: plan ahead for the full-scale exercise

Think about your organization or jurisdiction in relation to the special design considerations just discussed. Are there special problems you may need to work around when designing a full-scale exercise? Do you have ideas about how to make use of particular situations or locations? (Remember the description in Unit 1 of the building collapse exercise that was built around the planned destruction of a sports arena in Denver.)

In the space below, record notes that will help you remember your ideas about planning for a full-scale exercise in the future.

Site selection:

Scene management:

Personnel and resources:

Activity: Plan Ahead for the Full-Scale Exercise (Continued)

Response capability:

Safety:

Legal liability:

Emergency call-off:

The media:

Summary and transition

Unit 7 was the third of three units providing in-depth information about specific types of exercises. This unit provided information about the full-scale exercise, including key characteristics and special planning considerations. The next unit will focus on exercise evaluation.

Reference Library

For more information

- Domestic counterterrorism exercise "TOPOFF": www.usdoj.gov/opa/pr/2000/April/230ag.htm
- Lessons learnt from "TOPOFF": www.cdc.gov/ncidod/eid/vol6no6/hoffman.htm

Check

Knowledge check

Carefully read each question and all of the possible answers before selecting the most appropriate response for each test item. Circle the letter corresponding to the answer that you have chosen.

1. A full-scale exercise involves:

   a. All levels of personnel, including response personnel.
   b. Primarily the key policy-makers and decision-makers.
   c. Policy, coordination and operations personnel.
   d. The staff from one department or unit.

2. In a full-scale exercise, simulators:

   a. Are not required.
   b. Have a somewhat different role than in a functional exercise.
   c. Have the same role as in a functional exercise.
   d. Play the roles of all field personnel.

3. In a full-scale exercise the narrative is largely replaced by:

   a. Scripted messages.
   b. The exercise directive.
   c. A staged scene representing the emergency.
   d. The Player Handbook.

4. ______________ is a good use of a full-scale exercise.

   a. Testing emergency procedures and coordination of multiple agencies or organizations.
   b. Trying out a new and untested emergency plan.
   c. Training personnel in negotiation.
   d. Practising group problem solving in a non-threatening environment.

5. A full-scale exercise:

   a. Does not use pre-scripted messages.
   b. Uses only visual and action messages.
   c. May include visual, action, and pre-scripted messages.
   d. Uses the visual narrative in place of messages.

6. In a full-scale exercise, all decisions and actions by players occur in real time and generate real responses and consequences from other players.

   a. True
   b. False

Knowledge Check (Continued)

7. The best way to begin a full-scale exercise is to gather all of the participants in a central location for a briefing that explains the objectives, roles, responsibilities and ground rules.

   a. True
   b. False

8. _______________ is a major difference between full-scale and functional exercises.

   a. Formal evaluation
   b. Lead role of the controller
   c. Activation of the EOC
   d. Field-based action

9. Because field actions lead to decisions and further action in a full-scale exercise, it is unnecessary to develop a list of major and detailed events.

   a. True
   b. False

10. If you want to test response to a terrorist bomb attack, a good approach would be to stage an unannounced full-scale exercise at a sports arena during a professional game attended by thousands of spectators.

    a. True
    b. False

11. In designing a full-scale exercise, it is usually advisable to exclude the media and the public.

    a. True
    b. False

Answers

Knowledge Check (Continued)

1. a
2. b
3. c
4. a
5. c
6. a
7. b
8. d
9. b
10. b
11. b

Unit 8: Exercise Evaluation

UNIT 8

EXERCISE EVALUATION

Introduction

Evaluation is the process of observing and recording exercise activities, comparing the performance of the participants against the objectives, and identifying strengths and weaknesses.

Evaluation is a very complex topic, and this unit provides only a very general overview. In this unit, we will briefly discuss the importance of evaluation and its relationship to exercise development, evaluation team structure and duties, key aspects of evaluation methodology, and evaluation tasks that happen after the exercise is finished.

Unit 8 objectives

After completing this unit, you should be able to:

- describe the need for a systematic approach to exercise evaluation;
- identify and explain the tasks in the exercise evaluation process.

Why evaluate the exercise?

In order for an emergency management system to be effective, it is critical that the personnel, plans, procedures, facilities and equipment be exercised and tested on a regular basis. Yet no amount of exercising will be constructive unless each exercise is followed by a structured evaluation that enables the emergency management organization to identify successes and shortfalls.

Good evaluation can help the organization identify the following:

- Whether the exercise has achieved its objectives.
- Needed improvements in the EOP, procedures or guidelines.
- Needed improvements in the emergency management system.
- Training and staffing deficiencies.
- Needed operations equipment.
- Need for continued exercising of the plan and the emergency management functions.

If these goals are to be met, the evaluation approach must be systematic—methodical in procedure, thorough and organized.

Integrating evaluation throughout development

As you have seen in the previous units, exercise development is a continuous process which begins long before the exercise and continues after the exercise is over. It continues until needed changes have been made and suggestions have been incorporated into the next exercise.

Evaluation is not something that is done when all of the action is over. It begins when exercise design begins, when objectives are developed. You encountered the following chart in Unit 3. Notice again the kinds of evaluation activities that occur in the three exercise phases. In the remainder of this unit, we will take a brief look at each of the major evaluation activities listed in this chart.

Task categories			
	Pre-exercise phase	**Exercise phase**	**Post-exercise phase**
Design	▪ Review plan ▪ Assess capability ▪ Address costs and liabilities ▪ Gain support/issue exercise directive ▪ Organize design team ▪ Draw up a schedule ▪ Design exercise (8 design steps, including developing objectives)	▪ Prepare facility ▪ Assemble props and other enhancements ▪ Brief participants ▪ Conduct exercise	
Evaluation	▪ Select evaluation team leader ▪ Develop evaluation methodology ▪ Select and organize evaluation team ▪ Train evaluators	▪ Observe assigned objectives ▪ Document actions	▪ Assess achievement of objectives ▪ Participate in post-exercise meetings ▪ Prepare evaluation report ▪ Participate in follow up activities

The evaluation team

In the early stages of exercise design, a number of people will be brought together into a design team, led by a design team leader or exercise director. One member of this team will become the evaluation team leader, or chief evaluator, and this person will in turn select the members of the evaluation team.

Team structure

The size and composition of the evaluation team will depend on the type of exercise, its complexity, and the availability of people to serve. A small exercise with limited objectives, few participating organizations, and only a few locations might need only a team leader and three to six evaluators. A group this small would report directly to the team leader.

A large full-scale exercise might require an evaluation director to supervise several team leaders at different sites—who in turn supervise several evaluators. With this size team, the various evaluators would be assigned to serve at selected sites and a means of coordination or communication set up among them. An organizational chart would be helpful to keep lines of authority straight.

Role of the team leader

What? The evaluation team leader is primarily responsible for the evaluation methodology, for selecting and training the evaluation team, and for preparing the evaluation report.

Who? Ideally, the team leader should have experience in evaluation, management, exercise design or participation and training or education. Normally, he or she should be a member of the design team. In some cases, it might be necessary to bring in someone from outside the design team, although that is usually unnecessary if a volunteer training and exercise officer is on the design team and is willing to serve. (Other design team members are likely to be too heavily involved in developing the exercise.)

The Evaluation Team (Continued)

When? Selecting the evaluation team leader as early in the design process as possible has several advantages.

- It will ensure that the evaluation becomes an integral part of the exercise development effort.
- It will maintain the integrity of the evaluation function and prevent it from overlapping with the control and simulation functions.
- It will ensure that at least one person can devote time and mental effort to the large task of evaluation.

Selection of team members

The evaluation team leader is responsible for selecting and training the evaluation team. The ideal evaluator has many skills and personal attributes. Although it may not be possible to find a person who has all of these characteristics, the team leader will be looking for the following:

Skills	Attributes
▪ Appropriate technical expertise in evaluation ▪ Communication skills, both verbal and written ▪ Organizational ability ▪ Ability to see the relationship between events and objectives ▪ Ability to adjust to rapidly changing situations	▪ "People skills," sensitivity ▪ Objectivity ▪ Self-motivation ▪ Willingness to help ▪ Honesty and integrity (reports facts truthfully, keeps information confidential) ▪ Familiarity with the plan

Often an evaluation team can be assembled by using a little imagination and beating the bushes. Evaluators can be drawn from various sources.

- Neighbouring jurisdictions
- Emergency services personnel who will not be playing in the exercise
- Professional evaluators
- State or federal personnel
- College or university faculty
- Public service organizations

The Evaluation Team (Continued)

Training of team members

The training of an evaluation team usually can be done in an orientation meeting. The nature and length of the meeting will depend on the experience and skill of the members. Most evaluators—no matter how experienced—will need information on:

- the exercise scenario
- rules of play
- the objectives
- evaluation requirements and procedures
- evaluation forms.

Inexperienced evaluators may need some practice drills. Evaluators from outside the organization/jurisdiction will need information about the organization/jurisdiction.

> **Try to be unobtrusive**
>
> It is well documented that the presence of an evaluator can affect the behaviour of those being observed, possibly resulting in inaccurate data. Therefore, the evaluation team should plan ways to observe that are as unobtrusive as possible. Examples include:
>
> - being in position when the exercise begins so as not to attract attention;
> - postponing taking notes for a few minutes until players get involved in the play and stop noticing them.

Evaluation methodology

Evaluation methodology is simply the procedures and strategy used to evaluate an exercise. The methodology includes:

- how the evaluation team will be structured
- objectives to be measured
- evaluation packet.

Defining the team structure

Aspects of the team structure will have a significant impact on how the evaluation proceeds. Therefore, decisions such as the following are an important part of the methodology:

- The number of evaluators and their evaluation-related experience and knowledge;
- The organization of evaluators at multiple sites (i.e. sub-teams);
- Lines of authority (e.g. evaluators, supervisors or team leaders and chief evaluator);
- Communication and coordination among team members.

Evaluation criteria

One of the first steps in developing the methodology is to determine the criteria that will be used to determine if the exercise is successful. These criteria are tied to the objectives and expected actions.

At the outset of exercise design, the general objectives are defined. Then, in developing the scenario, the general objectives are broken into smaller units—the expected actions. From the expected actions, specific points of review and other evaluation measures can be developed.

Evaluation Methodology (Continued)

As discussed in Unit 4, the objectives must be stated clearly and precisely, describing actions that can be observed and measured. Using the SMART system will ensure that objectives are:

- **S**imple
- **M**easurable
- **A**chievable
- **R**ealistic
- **T**ask-oriented.

You may wish to review the material in Unit 4 now to refresh your memory about the important elements of good objectives.

Evaluation packet

The evaluation packet, or Evaluation Plan, contains all plans for the collection of data, including objectives and points of review, checklists or other evaluation forms, and observation techniques. A sample Evaluation Plan, with instructions for customizing it for a specific exercise, is provided in the "Samples" directory in the *Exercise Design Tool Box*, available at http://training.fema.gov/EMIWeb/IS/is139lst.asp.

Data can be collected by various means, such as evaluation forms, running written narrative, audiotape or videotape. Each method has advantages and disadvantages, which should be considered in developing the methodology. Some suggestions are on the next page.

Evaluation Methodology (Continued)

Suggested evaluation strategies

- **Plan the observation process.** One approach is to use these four steps.

  1. Recall the specific objectives of the exercise, the detailed events and the actions or decisions that they suggested.
  2. Identify the players expected to take the actions or make decisions as those who should be observed.
  3. Locate evaluators in a position to observe the players.
  4. Brief the evaluators on what actions or decisions are expected.

- **Provide points of review** to guide evaluators. They make it possible to be very objective in collecting data.

Example:

Objective. Demonstrate the adequacy of displays to support the emergency operations during the exercise.

Points of review:	YES	NO
▪ Status boards available in facility	______	______
▪ Status boards utilized	______	______
▪ Status boards kept updated	______	______
▪ Maps available	______	______
▪ Maps up to date	______	______

- **Provide evaluation forms.** They may include simple questionnaires, checklists, or rating sheets on which observations are recorded. They need not be complicated, but they must be objective, simple and specific.

Evaluation Methodology (Continued)

The following table illustrates a checklist for carrying out the four observation steps. (**Note:** This checklist also appears as Job Aid 19 in Appendix A.)

Sample observation checklist

Objective	Action/decision to look for	Players to observe	Where	Expected time
Notify principals	Activate "call-down" procedure	Superintendent	In EOC, school	10:15
Emergency contact of bus drivers	"Call-down" procedure—phone	Transportation Supervisor	EOC, School player	11:05
School closure announcements on TV/radio	Message preparation and distribution	Superintendent, Emergency Manager, Media	EOC, Media player	11:10
Open cafeteria and gym for shelter	Notify media, opening and supplying facilities	Superintendent, Facilities	EOC, School player	12:20

Key event monitoring

Most exercise scenarios include a number of events specifically designed to put stress on selected elements of the plan These are termed **key events**. Evaluators should pay special attention to these events.

> When a key event message is input, the evaluators monitor the participants' responses to the event. All responses should be noted on a Key Event Response Form. This form provides for multiple responses from several positions within the EOC as well as responses from outside the EOC.

Problem Log

The Problem Log allows participants, controllers, and simulators to document any observed action that may possibly create a problem. (While observing it should be understood by everyone that what may seem to be a plan or procedure problem may actually be a participant or simulator error.)

These potential problems can then be analyzed after the exercise to determine which ones are serious enough to require corrective action and to determine their source(s) —plan, preparedness, training or simulation.

Evaluation Methodology (Continued)

Sample forms

Examples of the following forms are provided on the next few pages. (**Note:** these forms also appear as Job Aid 20-25 in Appendix A.)

- Evaluator Checklist
- Narrative Summary Form
- Key Event Response Form
- Problem Log
- Exercise Debriefing Log
- Exercise Critique Form

Evaluator Checklist

Evaluator: ______________________________	**Date:** ________________
Location: _______________________________	

Objective No.:	**Function Being Evaluated:**

Objective:

Performance Criteria *[#]*

Points of Review:
Please answer the following: Y = Yes, N = No, NA = Not Applicable, NO = Not Observed

	Y	**N**	**NA**	**NO**
1.				
2.				
3.				
4.				

Comments:

____

____

____

____

____

____

____

____

____

____

Narrative Summary

Objective Number: ______________________ **Criterion Number:** ________________

Evaluator: ______________________________ **Location:** ______________________

Issue:
A specific statement of the problem, plan or procedure that was observed.

____
____
____
____
____
____
____

Discussion:
A discussion of the issue and its specific impact on operational capability.

____
____
____
____
____
____
____
____

Corrective Action Recommendation:
Recommended course(s) of action to improve performance or resolve the issue to improve operational capability.

____
____
____
____
____
____
____
____

Narrative Summary (Continued)

Office of Primary Responsibility:
The department, agency, or organization responsible for implementation of corrective actions.

Department, Agency, or Organization:
____

Individual Responsible: ____

Title: ______________ **Date Assigned:** ___ / ___ / ___ **Suspense Date:** ___ / ___ / ___

Key Event Response Form

Event No. ____________________ Scheduled Date/Time ____________________

Initially Input To ____________________ Actual Date/Time ____________________

Response Date/Time	Position Responding	Action Taken

Problem Log

Date: ______________________________

Exercise Assignment: ________________________________ Tel. No: _________________________

Time	Message Library No. (if known)	Problem	Analysis (Leave Blank)*

Exercise Debriefing Log

Exercise Debriefing Log

Exercise ____________ Recorder ____________ Date ____________

Problem Summary	Recommended Action	Responsible Agency/Person

Exercise Critique Form

Please take a few minutes to fill out this form. Your opinions and suggestions will help us prepare better exercises in the future.

1. Please rate the overall exercise on the following scale.

1	2	3	4	5	6	7	8	9	10
Very Poor								Very Good	

2. Compared to previous exercises, this one was:

1	2	3	4	5	6	7	8	9	10
Very Poor								Very Good	

3. Did the exercise effectively simulate the emergency environment and emergency response activities?
   Yes _________ No __________

   If no, briefly explain why:

4. Did the problems presented in the exercise adequately test readiness capability to implement the plan?
   Yes _________ No __________

   If no, briefly explain why:

Exercise Critique Form (Continued)

5. The following problems should be deleted or revised:

6. I suggest you add the following problems for the next exercise.

7. Please add any other comments or suggestions.

Post-exercise meetings

There are two types of post-exercise meetings, the player debriefing and meetings of the evaluation team.

Player debriefing

A short exercise debriefing should be conducted with the players immediately after the exercise. This debriefing gives them an opportunity to have their say about how things went, what they think should be changed, and commitments they might make.

This is how a debriefing generally is conducted.

- The **controller** conducts the debriefing, beginning with a review of broad objectives and commenting generally on both successes and shortfalls.
- The **controller** asks for a brief response (about 2 minutes each) from each player.
- As **participants** comment on their performance, the controller tries to maintain a balance between positive and negative comments and encourages everyone to contribute. Comments during the debriefing should be recorded for inclusion in the After Action Report. An Exercise Debriefing Log may be used for this purpose. An example is provided at the end of the unit.

Postexercise Meetings (Continued)

- The debriefing is for exercise participants. If **evaluators** wish to say a few words, they should concentrate on all the positive aspects of the exercise.

- It is a good idea to prepare a simple questionnaire for participants to fill out after the exercise. People who hesitate to enter into group discussions will often respond to a questionnaire. One possible format is a set of objective questions requiring only a check mark response, along with some open-ended questions about the performance in general, for example, "What was good or bad about the notification procedure?"

Keeping the debriefing on track

The purpose of the debriefing is to examine player performance. However, players will often want to critique the exercise itself: was it too long, too short, or had too many or too few messages. The goal is to keep the players on track, focusing on performance as much as possible. Explain that they will have an opportunity later to provide input into exercise design, then provide an exercise critique form for that purpose at the end of the debriefing. An example is provided at the end of this unit.

Evaluation team meetings

Evaluation team meetings are held to analyze the exercise and prepare the After Action Report. Evaluation team meetings may include the following:

- A meeting of the evaluation team(s) shortly after the exercise to compare notes;

- A more formal meeting of the team a week or so later to analyse the findings and develop an accurate account of what worked and what did not. The team analyses evaluation responses and any other data, and discusses how well each of the objectives was met;

- Additional meetings as needed to analyse data and prepare the report. The exercise design team may join the evaluation team at one or more meetings to offer feedback and suggestions. The report should be prepared within one to three weeks after the exercise, while memories are still fresh.

After Action Report

The findings of the evaluation team meetings are compiled in the After Action Report, which documents the effectiveness of the exercise. It serves as the basis for planning future exercises, upgrading the EOP or contingency plan, and taking corrective action.

The After Action Report is the responsibility of the evaluation team leader or chief evaluator, working with the evaluation team.

Form

The report may take a variety of forms. For example, a small exercise may warrant only a brief summary of the minutes of the player debriefing, followed by a few recommendations. Sometimes a memo will do the job. For large exercises—particularly functional or full-scale exercises—the report should be specific and comprehensive.

Format

There is no set organizational plan for an After Action Report. However, the topics listed in the following outline are usually covered.

After Action Report (Continued)

> **Sample report outline**
>
> - **Introduction**
>   (Main purpose of the report, why it is being submitted, preview of main topics, evaluation methodology used, and perhaps a general summary of main problems and recommendations)
>
> - **Statement of the problem**
>   (Purpose of the exercise)
>
> - **Exercise summary**
>   Goals and objectives
>   Pre-exercise activities
>   Participants and agencies
>   Description of exercise scenario
>
> - **Accomplishments and shortfalls**
>   Evaluation group findings
>   Summary of post-exercise debriefing
>
> - **Recommendations**
>   Training needs
>   Changes in the emergency plan
>   Other corrective actions

A copy of the After Action Report should go to the chief elected official of the jurisdiction (or head of the organization involved) and each participating entity.

Other reporting requirements

Other types of exercise reports are often required by agencies mandating exercises. Often they are simple checklists which require less time to prepare than a full written report. Check with the appropriate regulatory agencies for specific reporting requirements.

Implementing change

Recommendations for the future are the whole reason for conducting the exercise. The goals of an exercise are not achieved until the recommendations that come out of the evaluation are implemented. The purpose of the evaluation is to improve the emergency management plan and the organization's performance in carrying out that plan.

This is most likely to occur if objectives relate clearly to emergency functions and the focus of the evaluation is on performance, not people. Specifically, the change effort is centred around these issues:

- Are the procedures sound?
- Are resources sufficient to support the procedures?
- Are personnel adequately trained to follow the procedures and use resources?

Remember: we test plans, but we train people.

Activity

Activity: plan the evaluation

In Unit 4, you completed several design steps for a tabletop or functional exercise. In that unit, you:

- defined the scope
- wrote a statement of purpose
- developed three objectives
- outlined a narrative
- wrote two major events and two detailed events for each of them
- Composed a message.

Review the materials you developed in those activities, then complete the following questions with regard to the same exercise.

1. What size of evaluation team and what team structure would be appropriate for your exercise?

2. Who would you select for your evaluation team? What resources are available to you for obtaining qualified evaluators?

3. What kind of training would you provide for the evaluators?

Activity: Plan the Evaluation (Continued)

4. Briefly describe what evaluation methodology would be appropriate for your exercise. Consider:
   - observation strategies
   - documentation
   - forms
   - follow up

5. What type of report should be prepared, and how will it be generated?

6. How would you ensure that needed changes are implemented after the exercise?

Summary and transition

Unit 8 provided a general overview of the process for evaluating exercises. The next unit discusses materials and methods for enhancing an exercise simulation.

Check

Knowledge check

Carefully read each question and all of the possible answers before selecting the most appropriate response for each test item. Circle the letter corresponding to the answer that you have chosen.

1. Good evaluation can help the organization identify:

   a. Training and staffing deficiencies.
   b. Qualified members of the evaluation team.
   c. Exercise objectives.
   d. The scope of the exercise.

2. Evaluation begins when the exercise gets underway—usually when the narrative is presented.

   a. True
   b. False

3. It is usually best to keep the design team and the evaluation team separate and to bring in an outside evaluation team leader.

   a. True
   b. False

4. Every exercise should have an evaluation director, several teams of evaluators headed by team leaders, and a means of communicating among the teams.

   a. True
   b. False

5. The evaluation team leader is primarily responsible for the evaluation methodology, selecting evaluation team members, training the team, and preparing the evaluation report.

   a. True
   b. False

6. During the exercise, evaluators should:

   a. Ensure that players are aware of when they are being observed.
   b. Offer suggestions that can improve the exercise results.
   c. Focus on the positive.
   d. Avoid attracting players' attention.

Knowledge Check (Continued)

7. One thing evaluators should focus on during an exercise is:

    a. Explaining evaluation methodology to the players being observed.
    b. Having players explain their reasons for actions taken.
    c. Noting what actions are taken in response to key events.
    d. Finding as many positive points as negative ones.

8. A constructive evaluation strategy would be to have players, simulators, and controllers document problems they observe during the exercise.

    a. True
    b. False

9. The main purpose of the post-exercise debriefing is:

    a. To inform the participants of the evaluation results.
    b. To give the players a chance to comment on the exercise performance.
    c. To obtain feedback on the exercise design.
    d. To prepare the After Action Report.

10. The _________________ documents the effectiveness of the exercise and serves as the basis for taking corrective actions.

    a. Evaluation Plan
    b. Problem Log
    c. After Action Report
    d. Evaluator Checklist

Answers

Knowledge Check (Continued)

1. a
2. b
3. b
4. b
5. a
6. d
7. c
8. a
9. b
10. c

Unit 9: Exercise Enhancements

Unit 9

Exercise Enhancements

Introduction

Sometimes creativity is the difference between an adequate exercise design and one that really gets the participants excited and involved. This unit discusses the various ways that you can enhance an exercise with equipment, displays, people, props and other strategies.

Unit 9 objectives

After completing this unit, you should be able to:

- define the purpose of exercise enhancements
- identify resources available for exercise enhancement
- describe the benefits of specific exercise enhancement techniques.

Why use enhancements?

The point of an exercise is to simulate an emergency as realistically as possible. The more realistic the scenario, setting, atmosphere and equipment and materials made available to participants, the more likely players will fully engage with the action and get the most out of it.

A variety of exercise enhancements can help achieve this realism. In a drill or full-scale exercise, the use of real equipment and actual locations is inherently realistic. Added touches, such as simulated victims with convincing mock injuries, will make a functional or full-scale exercise even more realistic.

A tabletop or functional exercise must rely on materials and devices that you can bring into a room to increase the realism of the exercise.

It is not necessary to spend a lot of money or energy to enhance realism. Resources can include ordinary items that are currently available in the operations centre or command post, or common items that can be easily obtained.

Creativity

Creativity is the key to effective enhancements. There are many low-cost creative approaches you can try. Here are some examples.

- Videotape simulated "news broadcasts" depicting the disaster and taped interviews with "victims". Show these pre-recorded clips as part of the exercise.
- Audiotape news broadcasts and play them on the radio.
- Use make-up and props to simulate injuries.
- Use computers to chart plumes and provide data on river flows.
- If the telephone system is down as part of the exercise, leave the telephones in the EOC but don't connect them.
- If power is supposed to be out, then actually run a backup generator. Turn off the lights and computers if they're not hooked up to the generator. Of course this can disrupt the real office work flow, but it will result in a more realistic exercise.

These are just a few examples to get your creative juices flowing. We'll discuss more ideas in the remainder of the unit, and you will undoubtedly have ideas of your own. That is the whole idea!

Communications equipment

Various types of electronic equipment can be used to communicate the narrative and advance the scenario. Communications equipment can also be used to transmit some of the messages from simulators. In a highly simulated exercise, such as a complex functional exercise, try to transmit messages as you would in a real emergency— by phone, radio and TV.

> **Take advantage of what is available**
>
> There are always financial limitations, so plan to make use of the communications equipment your community normally has available during an emergency. Consider the following.
>
> - Landline telephone
> - Radio phones
> - Hotline dedicated phones
> - Portable/handheld radios
> - Cellular telephones
> - Military phone hook-ups
> - Citizens' band (CB)
> - Teletype systems
> - Fax machines
> - Amateur Radio Emergency Service (ARES)
> - Radio Amateur Civil Emergency Service (RACES)
> - Monitors/scanners
> - National Oceanic Atmospheric Administration (NOAA) Weather Radio
> - National Warning System (NAWAS)
> - Computerized Radio Packet

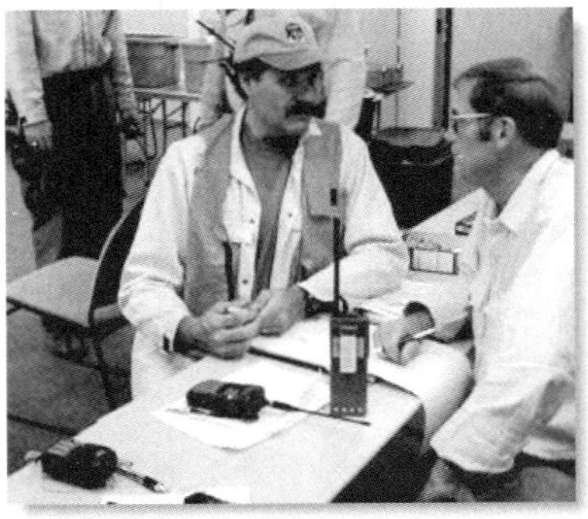

Visuals

No exercise is complete without a collection of visuals and displays, such as maps, charts, status boards, black or white boards, chart paper and easels. Videotapes and slides, although more difficult to obtain, can also enhance the realism of your exercise.

Maps

Uses. Maps provide context and detail for a scenario. Because they are essential to the handling of an actual disaster, they are necessary for all types of exercises. Even in an orientation or a tabletop exercise, maps provide useful information and give players a clearer picture of the simulated event. For example, they may be used to position equipment or to determine the nearest facilities for resource deployment.

Formats. Maps may be reproduced on paper for individual use or displayed on a wall. Acetate overlays make it possible to mark off areas or monitoring points and to reuse the map.

Types. The types and number of maps required depends on the exercise type and the hazard being tested. Below is a partial list of maps that you may find useful.

Types of maps	
▪ City street maps ▪ County street maps ▪ Subdivision maps ▪ Sewer maps (mains/facilities) ▪ Water maps (mains/facilities) ▪ Electric maps (lines/facilities) ▪ Gas maps (lines/facilities	▪ Flood plain maps ▪ Contour maps ▪ Police and fire district maps ▪ Centre city (downtown) maps ▪ Facility maps (e.g. plant layout, rail yards) ▪ Weather maps

Sources. Good maps can be obtained from a number of agencies. Before trying your hand as a cartographer, try local municipal agencies such as the City Planning Commission, Department of Highways, Engineering Department, Public Works. The U.S. Geological Survey (USGS) is another good source of maps related to earthquakes. FEMA's National Flood Insurance Programme has flood maps and NOAA has weather maps, satellite weather imagery and weather photographs.

Visuals (Continued)

Charts

The accumulation and sharing of information is an important operations centre function. To ensure coordinated and timely emergency response, visual displays allow everyone to quickly comprehend what actions have been taken and what resources and personnel are available.

Display needs will vary with the nature and scope of the exercise, but the charts listed below should be considered. Some displays are for use of the exercise staff only.

Type of chart	Description
Problem and Event Log	▪ Large events display board for posting major events. ▪ Should be available for all in the EOC to review, also useful for EOC shift change briefings. ▪ All major problems that are reported are entered on the log as they are received. ▪ May be divided into columns: nature of problem, problem number, assignment, response and remarks.
Damage Assessment Chart	▪ Divided into columns: areas reporting damage, time of report and extent of damage.
Facility Charts and Status Boards	▪ Used to track facilities involved in the exercise so participants are aware of available resources (companion maps are useful); examples of facility charts: • Hospitals: beds available, blood and other supply needs, personnel. • Congregate care facilities (e.g. those run by Red Cross or social service agencies): space available; status of food, water, bedding and medical stocks. • Law enforcement resources: numbers and locations of sworn, reserve, and auxiliary personnel; status of mutual aid units. • Fire resources: deployment and availability of fire units, status of fire mutual aid forms.
Organization Charts	▪ Useful for staff as a means of anticipating what agencies should be coordinating or reporting to other agencies. ▪ Optional.
Master Scenario of Events List	▪ Mainly for controller's use, to keep exercise on schedule, should not be seen by players. ▪ Contains detailed sequence of events developed as part of scenario.
Simulation Plotting Map	▪ Used by controller and simulators. ▪ Depicts pre-scripted input exercise information. ▪ Coded markers may be used to depict actions taken by various organizations (police, fire, medical/health, public works, utilities and Red Cross/voluntary agencies.

Visuals (Continued)

Computer resources

Many communities have their resources on a computerized inventory list. Where available, this inventory should be used during conduct of exercises to track resources.

Videotapes and slides

Videotapes can provide a very realistic presentation of an emergency which can be used to introduce the narrative or to give updates. News reports and interviews with politicians and the public can be pre-recorded to lend realism. Slides can be used for some of the same purposes.

Miscellaneous equipment and supplies

Sometimes exercise planners are so focused on the dramatic aspects of the exercise that they overlook common equipment and materials. Many of these items, available in most office stores at little cost, are indispensable.

Equipment and supplies to consider

- Projector for overhead transparencies or slides
- Copy machine
- Portable radio
- Pagers and cellular phones
- Public address system
- Pens, pencils and markers
- Chart paper
- Telephone books and directories
- Local and state contact lists
- EOC phone directory
- List of cellular phone numbers
- Name tags
- Resource lists
- Financial cost report form

People and props

In a full-scale exercise, the sense of reality occurs through the use of actual equipment in the actual setting. In this setting, fire trucks and the local airport would be considered types of enhancements.

Designers of full-scale exercises also bring in real people or props to enhance the realism. For example, an exercise simulating a hotel fire could use:

- fake smoke
- people playing the part of victims, made up to appear injured
- burnt boards and beams strategically placed at the event site
- a contained fire that the fire department would be required to extinguish.

Some jurisdictions use considerable ingenuity in these matters, creating model cities to use in tabletop exercises or obtaining mannequins to substitute for people in dangerous situations (e.g. trapped under a beam).

Such enhancements are limited only by budget, safety considerations and the imagination. So let your creativity work to enhance the exercise you are planning!

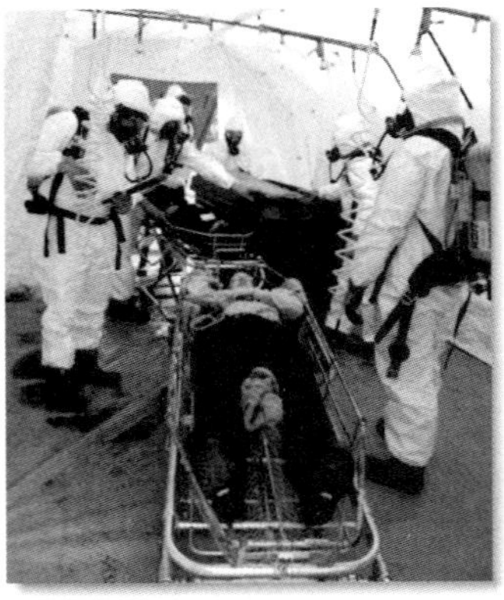

Enhancement resources

One of the problems designers often face is how to obtain materials, people and equipment that lend realism without breaking a very small budget. Consider soliciting volunteers and donations from other agencies in the community. Many are civic-minded and are happy to help out by lending equipment or providing volunteers to serve as victims. Below are some places to begin.

Potential resources	
▪ Hospitals ▪ Schools and colleges ▪ Business and industry ▪ Chemical Transportation Emergency Center (CHEMTREC) ▪ Boy and Girl Scouts ▪ Search and rescue volunteers ▪ Agencies ▪ Railroads ▪ The Salvation Army ▪ Public transportation	▪ Fire departments ▪ Police departments ▪ Chemical companies ▪ Amateur radio clubs ▪ Religious organizations ▪ National Guard/Military ▪ The Red Cross ▪ Service organizations (e.g. Elks, Lions, Rotary)

Costs and liability

In obtaining resources, it is important to consider costs (both initial and hidden) and the potential for liability.

- Person hours expended in obtaining and returning equipment or materials.
- Potential for damage or replacement costs.
- Arrangements for timely returns when items are borrowed or volunteers are "on loan." Too often "victims" have been left at the scene of the emergency site or the hospital because transportation plans were not made.
- Safety and liability concerns when using people and equipment. Be sure provisions are in place for liability insurance and equipment replacement.

Enhancement logistics

In planning for exercises, someone needs to take responsibility for managing the logistics related to enhancements. Be sure the following questions are answered.

- How will enhancements be used?
- Where will props be placed?
- Who will be in charge of props?
- How will people and props be picked up, transported and returned?
- What kind of return policy can be worked out for borrowed materials and equipment?
- In what condition must equipment be returned?
- Who will clean it?
- Is normal wear a concern?

Activity

Activity: enhance a scenario

Read the following scenario. What exercise enhancements could be used to add to the realism of this exercise? Write your ideas on the next page.

Functional exercise: potential airplane crash

Participating organizations: local dispatch, fire, police, search and rescue, EMS, emergency manager, mayor, airport crash/fire rescue, public works and hospital.

Location: your EOC or operations centre.

A jumbo jet that has experienced an inexplicable in-flight engine problem en route from Panama to New York will need to make an emergency landing at a large airport along the route. Though plans have been made to land at a city 200 miles north that is suitable for landing an aircraft this size, the latest communication with the pilot indicates that the plane has lost engine power and is losing altitude too quickly to reach the airport. Although the runway at your airport is too short to accommodate a jumbo jet, the only hope to save any of the 285 passengers and crew is to attempt a landing there.

Conditions at the airport are clear, but the surrounding area is very dry as a result of a continuing drought, and a hot, dry wind is blowing.

The airport is in a suburban area. However, the likelihood of the pilot being able to control the huge plane and land within runway limits is slim. The glide path passes over several high-density suburban housing developments.

The airport control tower has alerted its own crash/fire rescue (CFR) units and has requested that local emergency service units provide fire, police, medical and search and rescue assistance.

Garbled radio communication from the airliner alerts the airport control tower that the plane's hydraulic system is not functioning. The pilot's last message indicates that he will be attempting a soft-impact landing, but the plane breaks apart, crashing into the housing development and sending debris and bodies over an area of approximately one square mile. Smoke is visible in the area. A major part of the fuselage is at the beginning of the runway. CFR units are proceeding to the main crash site. All available resources are en route. However, traffic is at a standstill on the only roads leading to the site.

Activity: Enhance a Scenario (Continued)

What enhancements could be used to increase the realism of this scenario? Write your ideas below.

Possible enhancements

Activity: Enhance a Scenario (Continued)

Suggested answers

The following enhancements would contribute to the realism of this scenario. However, there are many ways to enhance an exercise, and you may have other ideas.

- City street maps.
- Subdivision maps for affected residential areas.
- Airport map.
- Fire district maps.
- Map/diagram of crash area, staging areas, command posts, ingress/egress routes (acetate overlay for updating movement of equipment and personnel are suggested).
- Resource status board showing available equipment (types, numbers) and assigned locations (computerized tracking if available).
- ICS organization chart.
- Damage assessment chart.
- Communications equipment to simulate communications among all involved agencies.
- Pre-recorded audiotape of pilot's transmissions to tower.
- Pre-recorded background noise audiotape of crash site (e.g. sirens, fire and voices) for atmosphere (could be accompanied by a pre-recorded video footage of firefighting, smoke).
- Master Scenario of Events List.
- Simulation plotting map.

Activity

Activity: plan enhancements for your exercise

Consider the exercise plans you developed in Unit 4. Answer the following questions about possible enhancements for that exercise.

1. Where will participants be located? Describe what kinds of facilities would be the most effective.

2. What kinds of communication equipment (if any) would be most appropriate for the exercise? What is available in your EOC or other operations centre? Where will you obtain what is not readily available?

3. Describe the visuals (maps, charts, status boards, computer resources, videotapes, slides and others) to be used, and where you can obtain them.

Activity: Plan Enhancements for Your Exercise (Continued)

4. How will you handle people and props?

5. What potential resources do you have for obtaining enhancements?

6. Who will be in charge of enhancement logistics?

7. What costs do you foresee associated with enhancements, and how will they be handled?

Summary and transition

Unit 9 provided information about the types of enhancements that can be added to increase the realism of an exercise, including communications equipment, maps, charts, display boards, computer resources, videotapes and slides, miscellaneous equipment and supplies, people and props. In the next unit you will have an opportunity to develop a functional exercise using the information and job aids provided throughout the course.

For more information

- NOAA National Climatic Data Centre:

  http://lwf.ncdc.noaa.gov/oa/ncdc.html

- NOAA Photo Library:

  http://www.photolib.noaa.gov/

- NOAA 3-D Weather Images:

  http://www.nnvl.noaa.gov/

Check **Knowledge check**

Carefully read each question and all of the possible answers before selecting the most appropriate response for each test item. Circle the letter corresponding to the answer that you have chosen.

1. The purpose of exercise enhancements is to:

   a. Increase credibility with the public and the media.
   b. Make the exercise more fun so more personnel will participate.
   c. Increase the realism so participants will respond as they would in a real event.
   d. Make it easier on the controller and the simulators.

2. In most tabletop and functional exercises, messages cannot be transmitted without sophisticated communications equipment.

   a. True
   b. False

3. Because they are essential to the handling of an actual emergency, maps are necessary for all types of exercises.

   a. True
   b. False

4. __________ are an important enhancement because they allow everyone to quickly comprehend what actions have been taken and what resources and personnel are available.

   a. Charts
   b. People and props
   c. Slides
   d. Audiotapes and videotapes

5. It is a good idea to seek out potential sources of materials, people and equipment in the local community to enhance an exercise.

   a. True
   b. False

Answers

Knowledge Check (Continued)

1. c
2. b
3. a
4. a
5. a

Unit 10: Designing a Functional Exercise

UNIT 10

DESIGNING A FUNCTIONAL EXERCISE

Introduction

In this unit you will put together all of the skills and knowledge you have learnt from previous units and develop a functional exercise. In Unit 4, you learnt the eight exercise design steps.

1. Assess needs.
2. Define scope.
3. Write a statement of purpose.
4. Define objectives.
5. Compose a narrative.
6. Write major and detailed events.
7. List expected actions.
8. Prepare messages.

You have already practised applying each of the design steps individually. Now you will put these steps together in one coordinated process. As an option, you may also develop the four major exercise documents (templates for which are found in the Exercise Design Tool Kit at http://training.fema.gov/emiweb/IS/is139lst.asp):

- Exercise Plan
- Control Plan
- Evaluation Plan
- Player Handbook

Unit 10 objectives

After completing this unit, you should be able to:

- design a small functional exercise using the eight-step design process.

Select your exercise

In Unit 2, you outlined a comprehensive exercise programme for your organization. Review that outline now. If you wish to revise or update the outline, you may do so.

Select a small functional exercise from your comprehensive plan. The exercise you choose should be fairly small, because of the time required to complete this course activity. If necessary, you may artificially limit a functional exercise in your plan.

If you used a functional exercise as the basis of your abbreviated design activities in Unit 4, you may either use the same exercise here (and develop it now in depth) or select a new one.

When you have selected your exercise, you may proceed with the design steps, as outlined in the remainder of this unit. As you undertake each design step, feel free to go back to earlier units and review the guidelines and examples that are provided.

Step 1: assess needs

The first step in developing any exercise is to assess the needs of your emergency management programme. Indicate below the most pressing problem areas in your community or organization. Base your assessment on your review of the emergency plans and procedures and what you have learnt from previous exercises or emergencies.

Exercise needs assessment

1. Hazards

List the various hazards in your community or organization. What risks are you most likely to face? You can use the following checklist as a starting point. **Note:** if your community has already conducted a hazard analysis, that is the best resource.

- ☐ Airplane crash
- ☐ Dam failure
- ☐ Drought
- ☐ Epidemic (biological attack)
- ☐ Earthquake
- ☐ Fire/urban conflagration
- ☐ Flood
- ☐ Hazardous material spill/release
- ☐ Hurricane
- ☐ Landslide/mudslide
- ☐ Mass fatality incident
- ☐ Radiological release
- ☐ Sustained power failure
- ☐ Terrorism
- ☐ Tornado
- ☐ Train derailment
- ☐ Tsunami
- ☐ Volcanic eruption
- ☐ Wildfire
- ☐ Winter storm
- ☐ Workplace violence
- ☐ Other ____________________
- ☐ Other ____________________
- ☐ Other ____________________

2. Secondary hazards

What secondary effects from those hazards are likely to impact your organization?

- ☐ Communication system breakdown
- ☐ Power outages
- ☐ Transportation blockages
- ☐ Business interruptions
- ☐ Mass evacuations/displaced population
- ☐ Overwhelmed medical/mortuary services
- ☐ Other ____________________
- ☐ Other ____________________
- ☐ Other ____________________
- ☐ Other ____________________
- ☐ Other ____________________

3. Hazard priority

What are the highest priority hazards? Consider the following factors.

- Frequency of occurrence.
- Relative likelihood of occurrence.
- Magnitude and intensity.
- Location (affecting critical areas or infrastructure).
- Spatial extent.
- Speed of onset and availability of warning.
- Potential severity of consequences to people, critical facilities, community functions and property.
- Potential cascading events (e.g. damage to chemical processing plant and dam failure).

#1 Priority hazard:

#2 Priority hazard:

#3 Priority hazard:

Step 1: Assess Needs (Continued)

4. Area

What geographic area(s) or facility locations are most vulnerable to the high priority hazards?

5. Plans and procedures

What plans and procedures—emergency response plan, contingency plan, operational plan, Standard Operating Procedures (SOPs)— will guide your organization's response to an emergency?

6. Functions

What emergency management functions are most in need of rehearsal? For example, what functions have not been exercised recently? Where have difficulties occurred in the past? You can use the following checklist as a starting point.

- ☐ Alert Notification (Emergency Response)
- ☐ Warning (Public)
- ☐ Communications
- ☐ Coordination and Control
- ☐ Emergency Public Information
- ☐ Damage Assessment
- ☐ Health and Medical
- ☐ Individual/Family Assistance
- ☐ Public Safety
- ☐ Public Works/Engineering
- ☐ Transportation
- ☐ Resource Management
- ☐ Continuity of Government or Operations
- ☐ Other ____________________
- ☐ Other ____________________
- ☐ Other ____________________

Step 1: Assess Needs (Continued)

7. Participants

Which agencies, departments, operational units and personnel need to participate in an exercise?

- Have any entities updated their plans and procedures?
- Have any changed policies or staff?
- Who is designated for emergency management responsibility in your plans and procedures?
- With whom does your organization need to coordinate in an emergency?
- What do your regulatory requirements call for?
- What personnel can you reasonably expect to devote to developing an exercise?

8. Programme areas

Mark the status of your emergency programme in these and other areas to identify those most in need of exercising.

	New	Updated	Exercised	Used in Emergency	N/A
Emergency plan					
Plan annex(es)					
Standard Operating Procedures					
Resource list					
Maps, displays					
Reporting requirements					
Notification procedures					
Mutual aid pacts					
Policy-making officials					
Coordinating personnel					
Operations staff					
Volunteer organizations					
EOC/Command Centre					
Communication facility					
Warning systems					
Utility emergency preparedness					
Industrial emergency preparedness					
Damage assessment techniques					

Step 1: Assess Needs (Continued)

9. Past exercises

If your organization has participated in exercises before, what did you learn from them, and what do the results indicate about future exercise needs? For example, consider the following questions:

- Who participated in the exercise, and who did not?
- To what extent were the exercise objectives achieved?
- What lessons were learnt?
- What problems were revealed, and what is needed to resolve them?
- What improvements were made following past exercises, and have they been tested?

Step 2: define the scope

Working from your needs assessment and the self-assessment you completed in Unit 3 about your organization's readiness for the exercise process, define the scope of the exercise (i.e. draw some meaningful limits). Make decisions according to (a) highest priority and (b) what can be realistically addressed in one exercise. Record your decisions below.

Scope worksheet

1. Highest priority hazards (major and secondary):

2. Geographic areas/locations of greatest vulnerability to these hazards:

3. Agencies/departments/organizational units: list below the entities that have a significant role in emergency management/response. Then enter check marks in any columns that apply.

Agency/Organization	Limited experience with major emergencies	New plans, staff, or organizational structure not yet exercised	Problems revealed in prior exercises

Step 2: Define the Scope (Continued)

4. Types/levels of personnel you want to have in the exercise:

- ☐ Policy-making (elected officials, chief operating officers, department heads)
- ☐ Coordination (managers, EOC representatives, department deputies)
- ☐ Operations (field personnel, headquarters staff level)
- ☐ Public representatives (media, PIOs, general public)
- ☐ Other: ______________________________

5. Types of operations/functions you want participants to engage in (e.g. notification, evacuation):

6. Degree of stress, complexity, time pressure the exercise should have:

	High	Medium	Low
Stress	________	________	________
Complexity	________	________	________
Time pressure	________	________	________

Step 2: Define the Scope (Continued)

Scope:
Type of emergency:
Location:
Functions:
Organizations and personnel:
Exercise type:

Step 3: write a statement of purpose

Develop your statement of purpose using the following template.

Statement of Purpose

The purpose of the proposed emergency management exercise is to improve the following emergency **operations**:

a. ______________________
b. ______________________
c. ______________________
d. ______________________

by involving the following **agencies/organizations/departments**:

a. ______________________
b. ______________________
c. ______________________
d. ______________________
e. ______________________
f. ______________________
g. ______________________
h. ______________________

in a ____________________ exercise simulating a ____________________
[type of exercise] **[type of emergency event]**

at ______________________________ on ________________.
[location] **[date]**

Step 4: define the objectives

In the space below, write objectives that support the purpose statement you composed in the last step. Remember the SMART system: Simple, Measurable, Achievable, Realistic, Task-oriented. Use extra paper if needed.

Objectives

Obj. No.	Objective	Organization

Step 5: compose a narrative

Outline the key points in your narrative using the following worksheet, then compose the script for the initial narrative in your exercise.

Narrative Outline
Event:
How fast, strong, deep, dangerous:
How you found out:
Response made:
Damage reported:
Sequence of events:
Current time:
Advance warning:
Location:

Step 5: Compose a Narrative (Continued)

Relevant weather conditions:
Other factors that would influence emergency procedures:
Predictions:

Step 5: Compose a Narrative (Continued)

Narrative Script

Step 6: write major and detailed events

In the space provided on the following pages, list events that might occur in your emergency scenario. Break down each of the major events into two or three detailed events.

Step 7: list expected actions

For each detailed event, write some actions that you would expect from participants involved in the exercise. Be sure to tie them to specific objectives (**note:** you may wish to number the detailed events and expected actions for later reference).

Step 7: List Expected Actions (Continued)

Events and Actions Planning Sheet

Obj. No.	Major Events	Detailed Events	Expected Actions	Organizations

Step 8: prepare messages

In the following message planning sheet, plan messages that will generate all of the expected actions you identified. Then compose the messages using the forms that follow.

Message Planning Sheet

Detailed Events	Expected Actions	Organizations	Message Outline

Step 8: Prepare Messages (Continued)

Emergency Exercise Message To: Method: From: Content:	**Emergency Exercise Message** To: Method: From: Content:
Emergency Exercise Message To: Method: From: Content:	**Emergency Exercise Message** To: Method: From: Content:

Step 8: Prepare Messages (Continued)

Emergency Exercise Message To: Method: From: Content:	**Emergency Exercise Message** To: Method: From: Content:
Emergency Exercise Message To: Method: From: Content:	**Emergency Exercise Message** To: Method: From: Content:

Step 8: Prepare Messages (Continued)

Emergency Exercise Message To: Method: From: Content:	**Emergency Exercise Message** To: Method: From: Content:
Emergency Exercise Message To: Method: From: Content:	**Emergency Exercise Message** To: Method: From: Content:

Step 8: Prepare Messages (Continued)

Emergency Exercise Message To: Method: From: Content:	**Emergency Exercise Message** To: Method: From: Content:
Emergency Exercise Message To: Method: From: Content:	**Emergency Exercise Message** To: Method: From: Content:

Constructing the Master Scenario of Events List

In the space below, develop a Master Scenario of Events List (MSEL) for the exercise you have just designed.

Sample Master Scenario Events List

Time	Message/Event	Expected Actions

Constructing the Master Scenario of Events List (Continued)

Sample Master Scenario Events List

Time	Message/Event	Expected Actions

Planning the exercise evaluation

In the space below, briefly describe your plan for evaluating the exercise.

Exercise Evaluation Plan

1. Evaluation team:

    - How many members and what team structure?

    - Who will you select, and where will you recruit them?

    - What kind of training will be provided?

Planning the Exercise Evaluation (Continued)

2. What methodology will be used to evaluate the exercise?

- Observation strategies:

- Documentation:

- Forms:

- Follow-up (What type of report? How will you ensure that needed changes are implemented?)

Planning the exercise enhancements

Using the following form, indicate your plans for facilities, displays, materials, equipment and other exercise enhancements.

Exercise Enhancements

Communications equipment:

- ☐ Hard-line telephone
- ☐ Radio phones
- ☐ Hot-line dedicated phones
- ☐ Portable/hand-held radios
- ☐ Cellular telephones
- ☐ Military phone hook-ups
- ☐ Citizens' Band (CB)
- ☐ Teletype systems
- ☐ Fax machines
- ☐ Amateur Radio Emergency Service (ARES)
- ☐ Radio Amateur Civil Emergency Service (RACES)
- ☐ Monitors/scanners
- ☐ NOAA Weather Radio
- ☐ National Warning System (NAWAS)
- ☐ Computerized Radio Packet

Maps:

- ☐ City street maps
- ☐ County street maps
- ☐ Subdivision maps
- ☐ Sewer maps (mains/facilities)
- ☐ Water maps (mains/facilities)
- ☐ Electric maps (lines/facilities)
- ☐ Gas maps (lines/facilities
- ☐ Flood plain maps
- ☐ Contour maps
- ☐ Police and fire district maps
- ☐ Centre city (downtown) maps
- ☐ Facility maps (e.g. plant layout, rail yards)
- ☐ Weather maps

Planning the Exercise Enhancements (Continued)

Charts:

- ☐ Problem and Event Log
- ☐ Damage assessment chart
- ☐ Facility charts
- ☐ Status boards (specify):
- ☐ Organization charts
- ☐ Simulation plotting map
- ☐ Other:

Other:

- ☐ Computer resources
- ☐ Videotapes
- ☐ Audiotapes
- ☐ Slides
- ☐ Projector
- ☐ Copy machine
- ☐ Portable radio
- ☐ Pagers and cellular phones
- ☐ Public address system
- ☐ Pens, pencils, and markers
- ☐ Chart paper
- ☐ Telephone books and directories
- ☐ Local and state contact lists
- ☐ EOC phone directory
- ☐ List of cellular phone numbers
- ☐ Name tags
- ☐ Resource lists
- ☐ Financial cost report form

Optional Activity: exercise design documents

Using the templates provided in the Exercise Design Tool Box (see the "Templates" directory), develop the following documents for the functional exercise you have just designed.

- Exercise Plan
- Control Plan
- Evaluation Plan
- Player Handbook

Summary and transition

Unit 10 provided an opportunity for you to apply the entire exercise design process in developing a functional exercise. Unit 11 will provide a brief course review and the final exam.

Unit 11: Course Summary

Introduction

This unit will briefly summarize the learning from the Exercise Design course. After completing this unit you should be able to summarize the key points of this course.

Reasons to exercise

Exercises are conducted in order to evaluate an organization's capability to execute one or more portions of its response plan or contingency plan. Exercises can be used to provide individual training and improve the emergency management system. Reasons to perform exercises include the following.

- Testing and evaluating plans, policies and procedures.
- Revealing planning weaknesses and resource gaps.
- Improve individual performance and organizational coordination and communications.
- Train personnel and clarify roles and responsibilities.
- Gain programme recognition.
- Satisfy regulatory requirements.

Comprehensive exercise programme

A comprehensive exercise programme is made up of progressively complex exercises, each one building on the previous one, until the exercises are as close to reality as possible. The programme must be carefully planned to achieve identified goals and should involve a wide range of organizations in its planning and execution.

Comprehensive Exercise Programme (Continued)

Types of exercises

There are five main types of activities in a comprehensive exercise programme.

- **Orientation seminar.** This is a low-stress, informal discussion in a group setting with little or no simulation. It is used to provide information and introduce people to policies, plans, and procedures.
- **Drill.** This is a coordinated, supervised exercise used to test a single specific operation or function. It involves deployment of equipment and personnel.
- **Tabletop exercise.** This is a facilitated group analysis of an emergency situation in an informal, stress-free environment. It is designed for examination of operational plans, problem identification and in-depth problem-solving.
- **Functional exercise.** This is a fully simulated interactive exercise that tests the capability of an organization to respond to a simulated event. It takes place in the EOC and focuses on coordination of multiple functions or organizations. Strives for realism, short of actual deployment of equipment and personnel.
- **Full-scale exercise.** This is a simulated emergency event, as close to reality as possible. It involves all emergency response functions and requires full deployment of equipment and personnel.

Building an exercise programme

Building an exercise programme is a multi-organizational team effort that includes:

- analysis of capabilities and costs
- goal setting
- development of a long-term plan
- scheduling of tasks
- public relations efforts.

The exercise process

The process of creating and staging an exercise includes a lengthy sequence of tasks that occur in three phases, before, during and after the exercise. Some of the tasks fall under the heading of design and some are part of evaluation.

Major task accomplishments

A simple way of viewing the exercise process is the sequence of five major task accomplishments:

1. Establishing the base.
2. Exercise development.
3. Exercise conduct.
4. Exercise critique and evaluation.
5. Exercise follow-up.

Establishing the base

Establishing the base is the process of laying the groundwork for the exercise. Some important aspects of this preparation are:

- reviewing the current plan
- assessing capability to conduct an exercise
- addressing costs and liabilities
- gaining support
- assembling and organizing a design team.

The Exercise Process (continued)

Exercise documents

Four major documents are developed during the exercise design process.

- Exercise Plan
- Control Plan
- Evaluation Plan
- Player Handbook

These documents are basically handbooks for particular audiences. Much of the content of these documents comes from the eight exercise design steps.

Exercise design steps

Tabletop, functional and full-scale exercises are based on a design process that includes eight steps.

1. Assess needs.
2. Define scope.
3. Write a statement of purpose.
4. Define objectives.
5. Compose a narrative.
6. Write major and detailed events.
7. List expected actions.
8. Prepare messages.

These are generally applicable steps regardless of the type of exercise. However, each type of exercise has some special considerations in how these steps are applied.

Exercise Design Steps (Continued)

Master Scenario Events List

Outputs from the design process are pulled together in the MSEL, a chart that the controller and simulators can use in keeping the exercise on track and on schedule.

The tabletop exercise

Exercise format

The tabletop exercise is essentially a group brainstorming session centred on a scenario narrative and problem statements or messages that are presented to members of the group. The format is informal, and the exercise is self-evaluated by the participants.

Facilitating a tabletop exercise

The facilitator is responsible for:

- setting the stage;
- distributing messages;
- stimulating discussion and pushing participants toward in-depth problem-solving;
- involving everyone;
- controlling and sustaining the action.

The Tabletop Exercise (Continued)

Designing a tabletop exercise

In applying the eight design steps, the first four steps (needs assessment, scope, purpose statement and objectives) are handled in the normal manner. The remaining steps can be simplified.

- The narrative can be relatively short.
- Only a few major or detailed events are required, and they are turned into problem statements.
- Expected actions must be identified, but they may involve such things as discussion or reaching consensus.
- Only a few (e.g. 10–15) carefully written messages or problem statements are needed.

The functional exercise

Exercise format

The functional exercise usually takes place in the operating centre and involves policy-makers and decision-makers. It uses an event scenario to test multiple functions or organizations, emphasizing coordination and communication. Participants include the following.

- Controller (the manager of the exercise).
- Players (people responding to the scenario within their normal roles).
- Simulators (people playing the parts of organizations and field units outside of the operations centre who deliver messages to players).
- Evaluators (observers who record actions taken in response to messages).

Participants respond in real time, adding an element of stress to the exercise. Communications equipment, displays and other enhancements can be used to add realism.

The Functional Exercise (Continued)

Controlling a functional exercise

The controller is responsible for:

- monitoring interaction and progress
- keeping the exercise on track
- dealing with the unexpected
- adjusting the pace as needed (the flow of messages can be adjusted by adding, deleting, misdirecting or reassigning messages).

Designing a functional exercise

The full eight-step process is used to develop functional exercises.

The full-scale exercise

Exercise format

The full-scale exercise combines the interactivity of the functional exercise with a field element and requires the coordination of the efforts of several organizations. It differs from a drill in that a drill focuses on a single operation and exercises only one organization.

The full-scale exercise achieves realism through:

- on-scene actions and decisions
- simulated victims
- search and rescue requirements
- communication devices
- equipment deployment
- actual resource and personnel allocation.

The Full-Scale Exercise (Continued)

Participant roles

All levels of personnel are involved, including policy-makers, coordination and operations personnel, and field personnel. A controller manages the exercise; volunteers simulate victims; evaluators observe and keep a log of significant events; and a safety officer ensures that potential safety issues are addressed.

Designing a full-scale exercise

After the first four design steps, the following special considerations apply to the design process.

- The narrative is largely a staged visual scene, so the written narrative can be minimal. The visual narrative must be planned in careful detail.
- Major and minor events are often presented visually and must be carefully planned.
- Expected actions must, as always, be specifically identified.
- Both visual and pre-scripted messages are used.

In a full-scale exercise, details are everything.

Site selection

The site selected for the event must have adequate space and be as realistic as possible without interfering with normal traffic or safety.

Scene management

Scene management involves planning and handling:

- logistics at the scene
- creation of a believable emergency scene
- number of victims
- management of props and materials
- number of controllers.

The Full-Scale Exercise (Continued)

Other special considerations

There are other special considerations in a full-scale exercise.

- Managing personnel and resources (many volunteers and lots of props).
- Ensuring that the emergency management system maintains response capability for routine events.
- Avoiding safety issues.
- Attending to issues of legal liability.
- Having a plan for emergency call-off.
- Working with the media.

Exercise evaluation

For an exercise to be useful, it must be accompanied by an evaluation—less formal for the tabletop exercise, but structured for the functional and full-scale exercises. Good evaluations can help the organization identify:

- whether the exercise has achieved its objectives;
- needed improvements in plans, procedures or guidelines, or the emergency management system as a whole;
- training and staffing deficiencies;
- equipment needs;
- the need for additional exercising.

The evaluation team leader, usually drawn from the design team, is responsible for evaluation methodology, selection and training of the evaluation team, and report preparation.

Exercise Evaluation (Continued)

Methodology

The evaluation methodology includes:

- evaluation team structure
- objectives to be measured
- evaluation packet (i.e. observation procedures and recording forms).

Post-exercise meetings

Post-exercise meetings include the player debriefing and meetings of the evaluation team to analyse the results and develop the After Action Report.

After Action Report

The After Action Report should describe the purpose of the exercise and address goals, objectives, pre-exercise activities, participants, scenario, accomplishments and shortfalls, and recommendations.

Exercise enhancements

Exercise enhancements are used to add to exercise realism. Depending on the type of exercise and available resources, enhancements may include:

- communications equipment
- visuals
- other equipment and materials
- people and props
- resources

Next steps

You have now completed IS139 and should be ready to take the final exam.

Complete the final exam in the back of the book by marking the correct responses.

Emergency Exercise

Job Aids

World Health Organization
Western Pacific Region

Adapted from the United States Federal Emergency Management Agency (FEMA)

Job Aids

Adapted from the United States Federal Emergency Management Agency (FEMA)

APPENDIX A

JOB AIDS

Job Aid 1: Exercise needs assessment

Use this tool to analyze where you may wish to focus your organization's exercise design efforts. In completing this needs assessment, you may wish to consult such resources as planning documents, demographic or agency data, maps, and training records.

1. Hazards

List the various hazards in your community or Organization. What risks are you most likely to face? For the purposes of this training workshop, we will use "Pandemic Influenza" as the hazard. You can use the following checklist as a starting point. **Note:** if your agency, community, state/province or nation has already conducted a hazard analysis and or impact analysis that is the best resource.

- ☐ Airplane crash
- ☐ Dam failure
- ☐ Drought
- ✓ Epidemic (Pandemic)
- ☐ Earthquake
- ☐ Fire/Firestorm
- ☐ Flood
- ☐ Hazardous material spill/release
- ☐ Hostage/Shooting
- ☐ Hurricane
- ☐ Landslide/Mudslide
- ☐ Mass fatality incident
- ☐ Radiological release
- ☐ Sustained power failure
- ☐ Terrorism
- ☐ Tornado
- ☐ Train derailment
- ☐ Tsunami
- ☐ Volcanic eruption
- ☐ Wildfire
- ☐ Winter storm
- ☐ Workplace violence
- ☐ Other ____________________
- ☐ Other ____________________
- ☐ Other ____________________
- ☐ Other ____________________

Job Aid 1: Exercise Needs Assessment (Continued)

2. Secondary hazards

What secondary effects from those hazards are likely to impact your organization?

- ☐ Communication system breakdown
- ☐ Power outages
- ☐ Transportation blockages
- ☐ Business interruptions – shortage of supplies
- ☐ Mass evacuations/displaced population
- ☐ Overwhelmed medical/mortuary services
- ☐ Other ____________________
- ☐ Other ____________________
- ☐ Other ____________________
- ☐ Other ____________________
- ☐ Other ____________________

3. Hazard priority

What are the highest priority hazards? Consider such factors as the following.

- Frequency of occurrence
- Relative likelihood of occurrence
- Magnitude and intensity
- Location (affecting critical areas or infrastructure)
- Spatial extent
- Speed of onset and availability of warning
- Potential severity of consequences to people, critical facilities, community functions and property
- Potential cascading events (loss of donor support)

#1 Priority hazard:

#2 Priority hazard:

#3 Priority hazard:

Job Aid 1: Exercise Needs Assessment (Continued)

4. Area

What geographic area(s) or facility location(s) is(are) most vulnerable to the high priority hazards?

5. Plans and procedures

What plans and procedures—emergency response plan, contingency plan, operational plan, standard operating procedures (SOPs)— will guide your organization's response to an emergency?

6. Functions

What emergency management functions are most in need of rehearsal? For example, what functions have not been exercised recently? Where have difficulties occurred in the past? You can use the following checklist as a starting point.

- ☐ Alert Notification (Emergency Response)
- ☐ Warning (Public)
- ☐ Communications
- ☐ Coordination and Control
- ☐ Emergency Public Information
- ☐ Damage Assessment
- ☐ Health and Medical (Acute PH)
- ☐ Individual/Family Assistance
- ☐ Public and Staff Safety
- ☐ Public Works/Engineering
- ☐ Transportation
- ☐ Resource Management
- ☐ Continuity of Operations (infrastructure, IT, Power)
- ☐ Other ______________________________
- ☐ Other ______________________________
- ☐ Other ______________________________

Job Aid 1: Exercise Needs Assessment (Continued)

7. Participants

Which agencies, departments, operational units and personnel need to participate in an exercise?

- Have any entities updated their plans and procedures?
- Have any changed policies or staff?
- Who is designated for emergency management responsibility in your plans and procedures?
- With whom does your organization need to coordinate in an emergency?
- What do your regulatory requirements call for?
- What personnel can you reasonably expect to devote to developing an exercise?

8. Programme areas

Mark the status of your emergency programme in these and other areas to identify those most in need of exercising.

	New	Updated	Exercised	Used in Emergency	N/A
Emergency Plan					
Plan Annex(es)					
Standard Operating Procedures					
Resource List					
Maps, Displays					
Reporting Requirements					
Notification Procedures					
Mutual Aid Pacts					
Policy-making Officials					
Coordinating Personnel					
Operations Staff					
Volunteer Organizations					
EOC/Command Centre					
Communication Facility					
Warning Systems					
Utility Emergency Preparedness					
Industrial Emergency Preparedness					
Damage Assessment Techniques					
Other:					

9. Past exercises

If your organization has participated in exercises before, what did you learn from them and what do the results indicate about future exercise needs? For example, consider the following questions.

- Who participated in the exercise, and who did not?
- To what extent were the exercise objectives achieved?
- What lessons were learnt?
- What problems were revealed and what is needed to resolve them?
- What improvements were made following past exercises and have they been tested?

Job Aid 2: Comprehensive Exercise Programme Planning Worksheet

Time frame:

Present problems:

Long-range goal:

Functional objectives:

Month:

Exercise:

For:

Purpose:

Rationale:

Month:

Exercise:

For:

Purpose:

Rationale:

Job Aid 2: Comprehensive Exercise Programme Planning Worksheet (Continued)

Month:

Exercise:

For:

Purpose:

Rationale:

Month:

Exercise:

For:

Purpose:

Rationale:

Month:

Exercise:

For:

Purpose:

Rationale:

Month:

Exercise:

For:

Purpose:

Rationale:

Job Aid 2: Comprehensive Exercise Programme Planning Worksheet (Continued)

Month:

Exercise:

For:

Purpose:

Rationale:

Month:

Exercise:

For:

Purpose:

Rationale:

Month:

Exercise:

For:

Purpose:

Rationale:

Month:

Exercise:

For:

Purpose:

Rationale:

Job Aid 3: Self-assessment—resources and costs

1. **Plans**

   How familiar are you with the emergency plans, policies, and procedures of your organization or jurisdiction?

   - ☐ Very familiar
   - ☐ Only general familiarity
   - ☐ Familiar with only a portion
   - ☐ Need to thoroughly review plans, policies, and procedures

2. **Time**

   a. How far in advance would your organization realistically have to schedule to plan and design each of the following exercise activities effectively?

   - Orientation ________________________________
   - Drill ________________________________
   - Tabletop exercise ________________________________
   - Functional exercise ________________________________
   - Full-scale exercise ________________________________

   b. How much preparation time can reasonably be allocated to developing an exercise?

   - Actual person days:
   - Elapsed time to exercise:

Job Aid 3: Self-Assessment: Resources and Costs (Continued)

Experience

When was your organization's last exercise?

a. What is your previous experience with exercises? Check all that apply.

Orientation:	☐ Presenter	☐ Participant		
Drill:	☐ Controller	☐ Participant		
Tabletop exercise:	☐ Facilitator	☐ Participant		
Functional exercise:	☐ Controller	☐ Simulator	☐ Player	☐ Evaluator
Full-scale exercise:	☐ Controller	☐ Responder	☐ Evaluator	☐ Victim

☐ Took part in post exercise debrief.

☐ Helped write an evaluation report.

b. What other exercise-related experience is available in your organization?

Facilities

What physical facilities do you use when conducting an emergency operation?

Will they be required for this exercise? Yes ☐ No ☐

Will they be available for this exercise? Yes ☐ No ☐

Job Aid 3: Self-Assessment: Resources and Costs (Continued)

Communications- what communication facilities and systems do you use in a real emergency?

Will they be required for this exercise? Yes ☐ No ☐

Will they be available for this exercise? Yes ☐ No ☐

Barriers: are there any resource barriers that need to be overcome to carry out this exercise?
Yes ☐ No ☐

If so, what are the barriers and how can they be overcome?

Costs

What types of costs might be incurred for these exercises in your organization? Do not list exact figures, just types of expenses, such as wages, salaries and transportation.

For an orientation:

For a drill:

For a tabletop exercise:

For a functional exercise:

For a full-scale exercise:

a. Are there ways that different organizations can reduce costs (e.g. by combining exercises, cost-sharing, resource-sharing) and still fulfil programme requirements? Explain.

Job Aid 4: Exercise Development Checklist

Mission
- ☐ ☐ Needs assessment
- ☐ ☐ Scope
- ☐ ☐ Statement of purpose
- ☐ ☐ Objectives

Personnel
- ☐ ☐ Design Team
- ☐ ☐ Controller or Facilitator
- ☐ ☐ Players
- ☐ ☐ Simulators
- ☐ ☐ Evaluators
- ☐ ☐ Management
  - ☐ Safety
  - ☐ Observers

Information
- ☐ ☐ Directives
- ☐ ☐ Media
- ☐ ☐ Public announcements
- ☐ ☐ Invitations
- ☐ ☐ Community support
- ☐ ☐ Management support
- ☐ ☐ Timeline requirements

Training/Briefings
- ☐ ☐ Train simulators, evaluators
- ☐ Controllers
- ☐ ☐ Players' Pre-exercise Briefing

Scenario
- ☐ ☐ Narrative
- ☐ ☐ Major/Detailed events
- ☐ ☐ Expected actions
- ☐ ☐ Messages

Logistics
- ☐ ☐ Safety
- ☐ ☐ Scheduling
- ☐ ☐ Rooms/Location
- ☐ ☐ Equipment
- ☐ ☐ Communications
  - ☐ Phones
  - ☐ Radio
  - ☐ Computers
- ☐ ☐ Enhancements
  - ☐ Maps
  - ☐ Charts
  - ☐ Other:

Evaluation
- ☐ ☐ Methodology
- ☐ ☐ Locations
- ☐ ☐ Evaluation forms
- ☐ ☐ Post-exercise debriefing

After Action Documentation/ Recommendations
- ☐ ☐ Evaluation meeting
- ☐ ☐ Evaluation report
- ☐ ☐ Follow-up ideas for next exercise

Job Aid 5: Activities Schedule

Deadline for Completion	Leader Activities	Team Activities

Job Aid 6: Design Team Worksheet

Name	Agency Represented	Contributions/Qualifications
Leader		
Members		

Job Aid 7: Scope Worksheet

1. Highest priority hazards (major and secondary): Will use Pandemic Influenza

2. Geographic areas/locations of greatest vulnerability to these hazards:

3. Agencies/departments/organizational units: list below the entities that have a significant role in emergency management/response. Then, enter check marks in any columns that apply.

Agency/Organization	Limited experience with major emergencies	New plans, staff, or organizational structure not yet exercised	Problems revealed in prior exercises

Job Aid 7: Scope Worksheet (Continued)

4. Types of operations/functions that you want participants to engage in (e.g. notification, display of information in operations room):

5. Degree of stress complexity, time pressure that the exercise should have:

	High	Medium	Low
Stress	________	________	________
Complexity	________	________	________
Time pressure	________	________	________

Job Aid 7: Scope Worksheet (Continued)

Exercise:
Scope:
Type of Emergency:
Location:
Functions:
Organizations and Personnel:
Exercise Type:

Job Aid 8: Statement of Purpose (Form 1)

The purpose of the proposed emergency management exercise is to improve the following emergency **operations:**

a. ____________________
b. ____________________
c. ____________________
d. ____________________

by involving the following **agencies/organizations/departments:**

a. ____________________
b. ____________________
c. ____________________
d. ____________________
e. ____________________
f. ____________________
g. ____________________
h. ____________________

in a _ **[type of exercise]** exercise simulating a _ **[type of emergency event]**

at ____________________ **[location]** on ____________________. **[date]**

Job Aid 8: Statement of Purpose (Form 2)

The purpose the proposed emergency management exercise is to:

- Coordinate the activities of health agencies, city and state/province government, volunteer organizations, and private industry in their response to a major incident;

- To provide training to staff;

- To test and evaluate the ____ Annexes;

- To enhance interagency coordination and cooperation by involving the following department or agency heads:

These entities will be tested on ______________________ in a simulated exercise involving a
[date]

______________________________________ at __________________________________
[type of incident] **[location]**

Job Aid 9: Objectives

List the exercise objectives below. Include the observable action, responsible party, conditions, and standards. Be sure each objective is SMART:

- **S**imple
- **M**easurable
- **A**chievable
- **R**ealistic
- **T**ask Oriented

Obj. No.	Objective	Organization

Job Aid 9: Objectives (Continued)

Obj. No.	Objective	Organization

Job Aid 9: Objectives (Continued)

Obj. No.	Objective	Organization

Job Aid 10: Narrative Outline

Event: **Novel Virus- Pandemic Imminent**
How fast, strong, deep, dangerous:
How you found out:
Response made:
Illness reported:
Sequence of events:
Current time:
Advance warning:
Location:
Other factors that would influence emergency procedures:
Predictions:

Job Aid 11: Events and Actions Planning Sheet

Events and Actions Planning Sheet

Obj. No.	Major Events	Detailed Events	Expected Actions	Organizations

Job Aid 12: Message Planning Sheet

Message Planning Sheet

Detailed Events	Expected Actions	Organizations	Message Outline

Job Aid 13: Emergency Exercise Message

EMERGENCY EXERCISE <MESSAGE>		
TO:	METHOD	FROM:
NO:	TIME:	

CONTENT: ____

____

____

____

____

ACTION TAKEN: ____

____

____

____

Job Aid 14: Sample Master Scenario Events List

Sample Master Scenario Events List		
Time	Message/Event	Expected Actions

Job Aid 15: Tabletop Exercise Checklist

Design

- ☐ Needs assessment, scope, statement of purpose, and objectives developed.
- ☐ Narrative:
  - ☐ May be shorter
  - ☐ Presented all at once or incrementally
- ☐ Events:
  - ☐ Limited number
  - ☐ Presented as problem statements
- ☐ Expected actions:
  - ☐ May involve identification of appropriate responses, identification of gaps in procedures, reaching group consensus, developing ideas for change, etc.
- ☐ Messages:
  - ☐ Limited number (e.g. 10–15)
  - ☐ Involve everyone
  - ☐ Tied to objectives

Facilitation

- ☐ Welcome participants
- ☐ Briefing:
  - ☐ Purpose and objectives
  - ☐ Ground rules and procedures
- ☐ Narrative presentation (printed, verbal, TV and radio)
- ☐ Ice breaker questions directed at high-ranking officers
- ☐ Messages organized to involve all organizations
- ☐ Strategies to encourage the reticent
- ☐ Facilitate—don't dominate
- ☐ Model positive behaviours (eye contact, positive reinforcement)
- ☐ Aim for in-depth problem solving
- ☐ Strategies for sustaining action
  - ☐ Multiple event stages
  - ☐ Varied pace
  - ☐ Balanced pace
  - ☐ Conflict resolution
  - ☐ Low-key atmosphere

Job Aid 16: Functional Exercise Message Flow Planning

	Participating Agency/Organization (List organizations above the columns below. Check the times when messages are scheduled for delivery to each Organization.)					
(Enter Msg.Times Below)						
Exercise Start						

Job Aid 17: Functional Exercise Design Checklist—Special Considerations

Facilities and equipment

- ☐ Sufficient work space for simulators and players
- ☐ Simulation room (if needed) near player room
- ☐ Space for message centre, control centre, observers (as needed)
- ☐ Clear work surfaces
- ☐ Communication equipment (telephones, switchboard)
- ☐ Parking
- ☐ Adequate ventilation and lighting
- ☐ Restrooms

Displays and materials

- ☐ Displays easily visible or accessible
- ☐ Maps (regional, state, local, area, downtown, operational units)
- ☐ Major events log, bulletin board, status boards, simulation plotting board
- ☐ Easels, chart paper
- ☐ Message forms
- ☐ Pencils/Paper
- ☐ Name cards

Beginning:

- ☐ "No-notice" or scheduled (according to objectives)

Briefing (short):

- ☐ Objectives
- ☐ Process
- ☐ Time period portrayed
- ☐ Ground rules and procedures

Narrative:

- ☐ Verbal, print, TV, computer, slides, or dramatization
- ☐ Time-skips if needed

Messages:

- ☐ Large number (depends on scope)
- ☐ Pre-scripted
- ☐ Optional pre-scripted for adjusting flow

Message delivery:

- ☐ Written
- ☐ Phone
- ☐ Other (verbal, speaker phone/radio, hand signals)
- ☐ Simulators prepared for spontaneous message development
- ☐ Standardized forms for written messages

Strategies for adjusting pace:

- ☐ Rescheduling
- ☐ Adding/Deleting messages

Job Aid 18: Full-Scale Exercise Planning Checklist—Special Considerations

Participants:
- ☐ Controller(s)—sufficient to manage all event sites
- ☐ Simulators (mock victims)—different age groups, body types, physical characteristics
- ☐ Players (most functions, all levels—policy, coordination, operation, field)
- ☐ Evaluators
- ☐ Safety Officer

Site selection:
- ☐ Adequate space for number of victims, responders, and observers
- ☐ Space for vehicles and equipment
- ☐ As realistic as possible without interfering with normal traffic or safety
- ☐ Credible scenario and location

Scene management:
- ☐ Logistics (who, what, where, how, when)
- ☐ Believable simulation of emergency
- ☐ Realistic victims
- ☐ Preparation of simulators to portray roles realistically
- ☐ Number of victims consistent with type of emergency, history of past events
- ☐ Types of injuries consistent with type of emergency, history of past events
- ☐ Victim load compatible with local capacity to handle
- ☐ Props and materials to simulate injuries, damage, other effects

Personnel and resources:
- ☐ Number of participants
- ☐ Number of volunteers for scene setup and victims
- ☐ Types and numbers of equipment
- ☐ Communications equipment
- ☐ Fuel for vehicles and equipment
- ☐ Materials and supplies
- ☐ Expenses identified (wages, overtime, fuel, materials and supplies)

Response capability:
- ☐ Sufficient personnel kept in reserve to handle routine non-exercise events

Safety:
- ☐ Safety addressed through development
- ☐ Each design team member responsible for safety in own discipline
- ☐ Hazards identified and resolved
- ☐ Safety addressed in pre-exercise briefing, simulator and evaluator packets
- ☐ Each field location examined for safety issues
- ☐ Safety officer designated, given authority

Legal liability:
- ☐ Legal questions of liability researched by local attorney

Emergency Call-Off_
- ☐ Call-off procedure in place, including code word/phrase
- ☐ Call-off procedure tested

Media:
- ☐ Role of media addressed in planning, used as a resource to gain favourable exposure
- ☐ Media and observers considered in logistical planning

Job Aid 19: Observation Checklist

Objective	Action/Decision to Look For	Players to Observe	Where	Expected Time

Job Aid 20: Evaluator Checklist

Evaluator: ______________________________	**Date:** ________________
Location: _______________________________	

Objective No.:	**Function Being Evaluated:**

Objective:

Performance Criteria *[#]*

Points of Review:
Please answer the following: Y = Yes, N = No, NA = Not Applicable, NO = Not Observed

	Y	N	NA	NO
1.				
2.				
3.				
4.				

Comments:

____

____

____

____

____

____

____

____

____

Job Aid 21: Narrative Summary

Objective Number: ______________________ **Criteria Number:** ________________

Evaluator: ______________________________ **Location:** ______________________

Issue:

A specific statement of the problem, plan, or procedure that was observed.

____
____
____
____
____
____
____

Discussion:

A discussion of the issue and its specific impact on operational capability.

____
____
____
____
____
____
____
____

Corrective Action Recommendation:

Recommended course(s) of action to improve performance or resolve the issue to improve operational capability.

____
____
____
____
____
____
____
____

Job Aid 21: Narrative Summary (Continued)

Office of Primary Responsibility:
The department, agency, or organization responsible for implementation of corrective actions.

Department, Agency, or Organization: ________________________________

Individual Responsible: ________________________________

Title: ________________ **Date Assigned:** ___ / ___ / ___ **Suspense Date:** ___ / ___ / ___

Job Aid 22: Key Event Response Form

Event No. ____________________ Scheduled Date/Time ____________________

Initially Input To ____________________ Actual Date/Time ____________________

Response Date/Time	Position Responding	Action Taken

Job Aid 23: Problem Log

Date: ____________________________

Exercise Assignment: ______________________________ Tel. No: ________________________

Time	Message Library No. (if known)	Problem	Analysis (leave blank)

Job Aid 24: Exercise Debriefing Log

Exercise Debriefing Log

Exercise ____________ Recorder ____________ Date ____________

Problem Summary	Recommended Action	Responsible Agency/Person

Job Aid 25: Exercise Critique Form

Please take a few minutes to fill out this form. Your opinions and suggestions will help us prepare better exercises in the future.

1. Please rate the overall exercise on the following scale.

1	2	3	4	5	6	7	8	9	10
Very Poor									Very Good

2. Compared to previous exercises, this one was.

1	2	3	4	5	6	7	8	9	10
Very Poor									Very Good

3. Did the exercise effectively simulate the emergency environment and emergency response activities?
   Yes _________ No __________

   If no, briefly explain why:

4. Did the problems presented in the exercise adequately test readiness capability to implement the plan?
   Yes _________ No __________

   If no, briefly explain why:

Job Aid 25: Exercise Critique Form (Continued)

5. The following problems should be deleted or revised:

6. I suggest that you add the following problems for the next exercise.

7. Please add any other comments or suggestions.

Appendix B: Acronym List

Appendix B Acronyms

ARES	Amateur Radio Emergency Services
CEO	Chief Elected Official
EAS	Emergency Alert System
EMERS	Emergency Management Exercise Reporting System
EMS	Emergency Medical Services
EOC	Emergency Operations Centre
EOP	Emergency Operations Plan
FEMA	Federal Emergency Management Agency
ICS	Incident Command System
JIC	Joint Information Centre
PIO	Public Information Officer
RC	Resident Coordinator
SOPs	Standard Operating Procedures
SAR	Search and Rescue
UN	United Nations
UNDSS	United Nations Department of Safety and Security
UNICEF	United Nations Children's Fund
WHO	World Health Organization

1. Research has shown that:

    a. Exercises are the best way to teach employees new skills.
    b. Extended lectures are an effective alternative to exercises.
    c. When possible, it is more effective to use real (rather than simulated) emergencies to test response procedures.
    d. People usually respond to an emergency in the way they have trained.

2. Which of the following is **not** a good reason to exercise?

    a. To reveal planning weaknesses and gaps in resources.
    b. To clarify roles and responsibilities.
    c. To reduce the need for organizational coordination and communication.
    d. To satisfy regulatory requirements.

3. In a progressive exercise programme, the exercises:

    a. Have a consistent format but are conducted with increasing frequency.
    b. Are organized to increase in complexity.
    c. Are organized to decrease in complexity.
    d. Are sponsored on a rotating basis by different organizations.

4. Availability of resources:

    a. Is irrelevant to the exercise process.
    b. Should not affect exercise design.
    c. Should be considered in scheduling and planning an exercise.
    d. Is seldom an issue for exercise design because exercises require few resources.

5. The major task accomplishments in the exercise process are:

    a. Establishing the base, exercise development, exercise conduct, critique and evaluation, and follow up.
    b. Conducting tabletop, functional, and full-scale exercises.
    c. Developing, conducting, and evaluating the exercise.
    d. Conducting a needs assessment, writing objectives, and evaluating whether the objectives were achieved.

6. The exercise design team leader should be:

   a. The Emergency Manager.
   b. Someone who is familiar with the emergency plan.
   c. A key operational member of a participating organization.
   d. The chief official of the jurisdiction or organization.

7. Which statement is true of an orientation?

   a. It requires field sites and actual equipment.
   b. It may be used to introduce or explain plans and policies.
   c. It involves a controller, simulators, and evaluators.
   d. It is used to test a specific operation.

8. Which statement is true of a drill?

   a. It is best conducted in a conference room.
   b. It involves a controller, simulators and evaluators.
   c. It is used to test a specific operation.
   d. It is aimed primarily at policy-makers and decision-makers.

9. Which statement is true of a tabletop exercise?

   a. It involves a highly realistic simulation.
   b. It involves a controller, simulators and evaluators.
   c. It requires field sites and actual equipment deployment.
   d. It is a facilitated analysis of an emergency situation.

10. Which statement is true of a functional exercise?

    a. It involves a controller, simulators and evaluators.
    b. It is simple, informal and stress-free.
    c. It requires field sites and actual equipment deployment.
    d. It may be used to introduce or explain plans and policies.

11. Which statement is true of a full-scale exercise?

    a. It is aimed primarily at policy-makers and decision-makers.
    b. It requires field sites but actual equipment remains in the shed.
    c. It is used to test a specific operation.
    d. It involves a highly realistic simulation.

12. The narrative:

   a. Sets the stage for later action.
   b. Lists all of the events that will occur in the exercise.
   c. Provides the master list of events and expected actions.
   d. Explains the objectives of the exercise.

13. The scope:

   a. Sets the mood for the exercise.
   b. Establishes the limits of the exercise.
   c. States the objectives of the exercise.
   d. Lists the financial and human resources required to run the exercise.

14. An example of a good exercise objective is:

   a. A sufficient number of accurate messages will be transmitted by the dispatch centre to the communications centre and primary response organizations in a timely and efficient manner.
   b. At the time the evacuation notice is received, the EOC policy and coordination groups will examine the needs of schools and other special facilities and organize notification according to SOPs.
   c. Proper procedures will be followed to declare a disaster or ask for outside aid.
   d. Interaction with other jurisdictions will be demonstrated.

15. A ________________ is a useful tool to help the controller keep the exercise on track and on schedule.

   a. Message form
   b. Narrative
   c. Master scenario of events list
   d. Exercise directive

16. A convincing, unified scenario requires:

   a. 10 major events and approximately 100 detailed events.
   b. Messages written in a way will not allow unexpected responses to occur.
   c. The players know the events in advance.
   d. Careful scripting of events.

17. In a tabletop exercise, an important part of the facilitator's job is to:

    a. Maintain an even pace and consistent approach.
    b. Adhere to a highly structured agenda.
    c. Sustain action and keep everyone involved.
    d. Make sure the entire set of problem statements is discussed.

18. Which of the following is **not** true of a simulator in a functional exercise:

    a. May deliver written messages.
    b. Is often called upon to rate the performance of key players.
    c. May deliver messages verbally.
    d. Sometimes needs to make up a response to a player.

19. An exercise that tested only notification procedures in response to a terrorist bombing would be:

    a. A drill.
    b. A tabletop exercise.
    c. A functional exercise.
    d. A full-scale exercise.

20. ______________ is a good use of a full-scale exercise:

    a. Trying out a new and untested emergency plan.
    b. Training personnel in negotiation.
    c. Practicing group problem solving in a non-threatening environment.
    d. Testing emergency procedures and coordination of multiple agencies or organizations.

21. A full-scale exercise involves:

    a. All levels of personnel, including response personnel.
    b. Primarily the key policy-makers and decision-makers.
    c. Policy, coordination, and operations personnel.
    d. The staff from one department or unit.

22. A good evaluation can help the organization identify:

    a. Training and staffing deficiencies.
    b. Qualified members of the evaluation team.
    c. Exercise objectives.
    d. The scope of the exercise.

23. During the exercise, evaluators should:

   a. Ensure that players are aware when they are being observed.
   b. Offer suggestions that can improve the exercise results.
   c. Focus on the positive.
   d. Avoid attracting players' attention.

24. One thing evaluators should focus on during an exercise is:

   a. Explaining evaluation methodology to the players being observed.
   b. Having players explain their reasons for actions taken.
   c. Noting what actions are taken in response to key events.
   d. Finding as many positive points as negative ones.

25. The purpose of exercise enhancements is to:

   a. Increase credibility with the public and the media.
   b. Make the exercise more fun so more personnel will participate.
   c. Increase the realism so participants will respond as they would in a real event.
   d. Make it easier on the controller and the simulators.